Conifer Reproductive Biology

Claire G. Williams

Conifer Reproductive Biology

Springer

Claire G. Williams
USA

ISBN: 978-90-481-8167-4 e-ISBN: 978-1-4020-9602-0
DOI: 10.1007/978-1-4020-9602-0
Springer Dordrecht Heidelberg London New York

Cover Image: Snow and pendant cones on spruce tree (reproduced with permission of Photos.com).

Printed on acid-free paper

Springer is part of Springer Science+Business Media (www.springer.com)

Foreword

When it comes to reproduction, gymnosperms are deeply weird. Cycads and conifers have drawn out reproduction: at least 13 genera take over a year from pollination to fertilization. Since they don't apparently have any selection mechanism by which to discriminate among pollen tubes prior to fertilization, it is natural to wonder why such a delay in reproduction is necessary. Claire Williams' book celebrates such oddities of conifer reproduction. She has written a book that turns the context of many of these reproductive quirks into deeper questions concerning evolution. The origins of some of these questions can be traced back Wilhelm Hofmeister's 1851 book, which detailed the revolutionary idea of alternation of generations. This alternation between diploid and haploid generations was eventually to become one of the key unifying ideas in plant evolution. Dr. Williams points out that alternation of generations in conifers shows strong divergence in the evolution of male and female gametes, as well as in the synchronicity of male and female gamete development. How are these coordinated to achieve fertilization?

Books on conifer reproduction are all too rare. The only major work in the last generation was Hardev Singh's 1978 *Embryology of Gymnosperms,* a book that summarized the previous century's work. Being a book primarily about embryology, it stopped short of putting conifer reproduction in a genetic or evolutionary context. There have also been reviews in particular orders and families, but these tend to be oriented towards orchard management problems and practical concerns of seed production. There are few works that are comprehensive and yet accessible to graduate students and researchers. Professor Williams' book provides new avenues to explore in our thinking.

The book also comes along at a opportune moment. The last 40 years have witnessed many advances in our understanding of conifer reproduction. These have proven of practical value to both experimentalists and seed orchard managers. Advanced breeding programs have developed in many countries to the benefit of the forest industry. Genetically improved trees are widely planted, providing sources of fiber and timber. In addition, these breeding methodologies have spread to many developing countries. Although we are only beginning to understand some of the cell biology that controls and regulates the properties of economically important traits in trees, we have developed many practical and effective methodologies that allow us to produce seed at scale for purposes of reforestation and afforestation.

With such an abundance of genotypes selected and tested for various characteristics, it is not surprising that conifer reproduction studies are poised to enter another era. But what era is it? Conifer genomics and metabolomics are expanding; conifer proteomics is still in its infancy. It appears that the original sin of conifers is to have too much DNA. We are forced to be wait for a published genome. This is compounded by our inability to see where things are going in the future. Arabidopsis, poplar and other model plants are providing new information on herbs and trees. Turning this to our advantage rests on our scientific imagination. In this regard, Dr. Williams' book takes stock of interesting questions for the benefit of future forest biologists and geneticists.

The book works at a number of levels. It introduces conifers in their proper evolutionary context. It begins by putting a modern relic – Wollemi Pine – in perspective. From the Golden Age of conifers in the Mesozoic when they were the world's first truly successful and widespread seed plants, Claire Williams leads us to modern conifers, including those that are rare and threatened. She then turns to reproduction, in particular of the Pinaceae. She sets the stage by describing the Bauplan of the life cycle from meiosis to mature seed. The book then changes gear when it considers the consequences of heterospory. This is of great value for training graduate students and for providing researchers with new questions concerning karyotypes, large genomes, and the consequences of crossing. Generally, in the previous compendia of conifer reproduction, emphasis has been on family-by-family differences in the details of reproduction. Not since Willson and Burley's 1983 book entitled *Mate Choice in Plants* has there been a book in which the effects of conifer organization on reproduction and genetics are presented as unsolved problems.

From the historical perspective, many things remain as true today as when they were first unveiled. As the author writes, "Robert Brown's (1828) discovery of the gametophyte and its multiple archegonia are more profound than he knew". We still have little idea how these large female gametophytes regulate their own lives separately from those of the surrounding sporophyte, whether in terms of defense or signaling or regulation of development. Furthermore, we have little understanding of the female gametophyte's role in regulating its own abortion or that of young embryos. The male side of reproduction poses other problems, starting with pollen dispersal. The description of Erdtman's trip across the Atlantic in the 1930s during which he sampled the air for pollen provides a nice context in which to consider pollen dispersal. In this chapter, we get a fresh consideration of features that allow pine pollen to fly long distances.

Synchronizing pollination, female development and fertilization in these reproductive plodders remains a great mystery. Claire Williams lays it out carefully, providing many new and testable ideas. What are the forces that converge to synchronize male and female development and result in pollination drop secretion? How does the female gametophyte eventually exert its influence in reproduction, given that for so long it was subsidiary to the sporophyte in all previous pollination-related events? Her years of study of events that regulate embryo development and death, particularly those due to selfing, show clearly in the later chapters on seed development and mating system dynamics, culminating in a thought-provoking discussion of embryo lethal systems.

This is a book that should be used by researchers who want new perspectives on conifer reproduction. It is an excellent book for a graduate course or to put new life into a discussion group or journal club. Since the 1970s, there has been an explosion in our understanding of the details of reproduction, but it is now time to take stock and think about what this means. This book will help young and old researchers frame new questions, re-frame old ones using new genomic tools, and finally, it encourages readers to ask good questions that will contribute to our knowledge of the evolution, ecology and development of conifers.

Victoria, BC, Canada Patrick von Aderkas

Prologue

This book on conifer reproductive biology is intended as a text supplement for plant biology courses. Knowledge of model flowering plants is expanding so fast that each new plant biology text has less written on conifers than the last. *Conifer Reproductive Biology* seems needed as a specialized botany reference for life science professionals, graduate students and advanced undergraduates. Its content, chosen for its relevance to those working in the life sciences including ecology, evolution, genomics, environmental sciences, genetics, forestry, conservation and even immunology, is shaped by a trend towards the integrative study of conifer reproduction.

Book content. Integrative reproductive biology rests on a concept put forth by Johann Goethe (1749–1832) who argued that each of the various parts of a plant – including its reproductive organs and seeds – could be understood as variations on a basic body plan. His *Bauplan* concept for the Die Urpflanze ("the original plant")[1] enables one to predict selection space for plant morphology or to infer plant forms no longer living. Although the *Bauplan* concept was first adopted as a central tenet in metazoan biology, the seed plant biology community is only now adopting this useful concept. One must consider the possibility that conifers might not share a *Bauplan* distinct from other seed plants. After all, the grouping of conifers is a taxonomic classification based on the *absence* of a carpel and other angiosperm reproductive characters.

An integrative biology perspective is provided here for the few conifer lineages that survived after the Mesozoic era. Conifers and other gymnosperms are the only living links to early seed plant reproduction so *Conifer Reproductive Biology* has evolutionary relevance. This perspective includes the *Bauplan* concept. It will come into sharper focus now once the huge conifer genomes are completely sequenced but until then, seed plant biology (inclusive of conifers) will continue to move through a reductionist phase where whole-organism studies are less valued. But such reductionism is only temporary. Next will be the need to integrate all sources of information,

[1] The *Bauplan* concept was published in 1790 by Johann Wolfgang von Goethe in his *Metamorphose der Pflanzen* as cited by Kaplan, D. (2001). The science of plant morphology: definition, history, and role in modern biology. American Journal of Botany 88: 1711–1741.

from phenotypic data to DNA code so this book should provide a foundation for conifer reproduction relevant to integrative hypothesis construction.

Two other ideas shaped the book's content. First is the question of extinction. Although they are Mesozoic relics, conifers as a group are not going extinct. A few are aggressive colonizing species such as *Pinus taeda* which are actually favored by human disturbance but other species face a high risk of extinction requiring intervention. The book's anecdotes focus more on the latter group but the reader should not take away the idea that all conifers are going extinct.

The second idea is that conifer reproduction research has been historically funded in the twentieth century because of its commodity significance. Three or four genera within the Pinaceae have been more closely studied than all of other conifers and gymnosperms put together. It is unfortunate that such uneven research funding skews this book's content and organization; more basic research is sorely needed.

Book organization. The book is framed around the diplohaplontic life cycle. Section I provides an overview of conifers and their life cycle. Section II presents each of the different working parts to the mating system. Divergence – and yet synchrony – of female and male reproductive structures is the basic message. Here it also becomes apparent that this book's title should really be "*Conifer Reproductive Biology: With Emphasis on the Pinaceae*" because here each topic starts with what is known about *Pinus*, moves to the Pinaceae and the Cupressaceae and then finally ends with functional and morphological variants observed in other conifers. In a few cases, *Ginkgo biloba* and a few other gymnosperms are mentioned.

The reader should note that Hardev Singh's book *Embryology of Gymnosperms* provides a far more comprehensive treatment of gymnosperm reproduction, particularly for those taxa indigenous to the Southern Hemisphere. Section III shows the sum of the working parts: how this dynamic mating system relies more on form than chance to produce a viable seed. The last chapter is the special case of how selfing and the embryo lethal system shape mate choice in a sessile life form.

Available literature on conifer reproduction is vast, spanning more than 3 centuries so all contributions could not be included in this book. The reader is encouraged to view this book as only a starting place for reading all of the original literature on a given topic.

History for conifer reproduction worldwide goes well beyond scientific contributions or even recorded history. Humans have relied on conifer forests for food, medicine, shelter, adornment, travel and warfare for millennia. Such value extends to reproductive characters too. Edible pine seeds have been part of the human diet – and even sacred rites – for thousands of years.[2] Even today, pine pollen from Asia is bottled and sold as an anti-aging health tonic[3] a practice which contrasts with

[2]Zach B. (2002) Vegetable offerings on the Roman sacrificial site in Mainz Germany – short report on the first results. Vegetation History and Archaebotany 11: 101–106. Also see Mirov 1967 *The Genus Pinus*. Ronald Press, New York. 602 p. and The Gymnosperm Database at http://www.botanik.uni-bonn.de/conifers/topics/index.htm

[3]Yunamite® pine pollen beverage is a pure and natural pine product from the mountains of Yunnan, China http://www.pharmeast.com/id8.htm 11/11/02

pollen from a few members of the Cupressaceae[4] which induces a medical condition known as cedar fever. Conifer reproduction has been the subject of mythology, art and religious practices and this continues unabated. Today, conifers still inspire poetry, state flags, township names and postage stamps. The book includes a few anecdotes which go to this point but the reader is encouraged to explore further.

Acknowledgements. Composing this book would not have been possible without the resources of many world-class library collections: Oxford University's Radcliffe Library; Harvard University's Botany libraries and the Arnold Arboretum Horticultural library, Duke University's Biology and Environmental Sciences Library and the University of British Columbia's Life Science Libraries.

Primary financial support came from the John S. Guggenheim Foundation. Other sponsors include the U.S. Forest Service's Southern Forest Experiment Station, the Canada–U.S. Fulbright Program, Harvard's Bullard Fellows Program, Oxford University's Christ Church College, NATO, Winrock International, University of Victoria's Forest Biology Centre, British Columbia's Ministry of Forestry, American Association for Advancement of Science, U.S. State Department and Vietnam National University in Ho Chi Minh City. Completion of this book would not have been possible without this support nor without the tenured full professorship once held at Texas A&M University.

My special thanks are due to Floyd Bridgwater, USDA-Forest Service (retired) for his photographs, protocols and comprehensive editorial efforts. I want to thank other reviewers: Patrick von Aderkas, Thomas Blush, Eric Brenner, William Carlson, Gabriel Katul, Patricia Gensel, Michael Greenwood, James Grob, Ben LePage, Anne Raymond, David Remington, John Russell, Frank Sorensen and Barry Tomlinson. Even though their corrections have been numerous, I accept full responsibility for the book's content. I also thank anonymous reviewers acting on behalf of the following scientific journals: *Canadian Journal of Botany*, *Canadian Journal of Forest Research*, *Forest Ecology and Management*, *Genetics*, *Genome*, *Heredity*, *Nature Biotechnology*, *New Phytologist* and *Molecular Ecology*.

January 24, 2009 Claire Williams

[4] Canini A. et al. 2004. Localisation of a carbohydrate epitope recognised by human IgE in pollen of Cupressaceae. Journal of Plant Research 117: 147.

Contents

Section I
Conifer Reproductive Biology Overview

Plate I Shown here is the fossil remains of a *Larix* cone preserved at high-latitude middle Eocene deposits found on Axel Heiberg Island, Nunavut Territory Canada. Many genera of the Pinaceae including this *Larix* species were part of these *Metasequoia*-dominated forests that grew north of the Arctic Circle at paleo-latitudes 75–80°C more than 45 million years ago. Conifers were once more prevalent than they are today (Photograph by Ben LePage. Permission granted)

Chapter 1
Introducing Conifers

Summary Conifers are cone-bearing seed plants with an ancient evolutionary history. Opening with an introduction to Australia's Wollemi pine, one finds that modern conifer taxa (seven families, 71 genera, 620+ species) are persistent Mesozoic relics. As such, their evolutionary history begins with the terrestrial invasion of land plants and the greening of the earth, the rise of the Paleozoic forest and the Jurassic plant diet of herbivorous dinosaurs. All modern conifers, not just Wollemi pine (*Wollemia nobilis*), are living fossils. Conifers have persisted despite continental drift, climate oscillations, volcanism and the rapid spread of angiosperms. Modern conifers, as a whole, are distributed worldwide although a few regions of the world such as China, Mexico and New Caledonia have high concentrations of conifer taxa. Although many conifer species have large, wide-ranging census populations, others such Wollemi pine are critically endangered. Vestiges of the ancient conifer diaspora can be seen in the fossilized *Metasequoia*-dominated forests in Canadian High Arctic and from the endemic Da Lat ecosystem in Vietnam which includes the flat-leaved *Pinus krempfii*. Conifers are among the oldest extant seed plant lineage and their peculiar reproductive biology holds clues about seed plant evolution.

In 1989, the author hiked extensively through Wollemi National Park in New South Wales, Australia where she unknowingly walked past a discovery of a lifetime – a hidden stand of Wollemi pine. Five years later, this same stand of trees would be identified as the living fossil *Wollemia nobilis* and heralded as the discovery of the century. This new conifer species, a close match to araucaroid fossils and thus long thought to be extinct, was a remnant from ancient Gondwana forests in the Southern Hemisphere. This find was truly a serendipitous treasure for science. Fewer than 100 individuals exist today so preserving the treasure meant propagating the tree into several botanical gardens. Today *Wollemi nobilis* is more than a botanical curiosity; this tree is a critical link to elucidating seed plant evolution. What follows next is what the discovery of Wollemi pine actually means to the larger evolutionary history of conifers.

Conifers, including the Pinaceae, are gymnosperms or "naked seeds." As the name implies, gymnosperms are the earliest extant seed plant lineages, having

C.G. Williams, *Conifer Reproductive Biology*,

evolved many millions of years ago. Conifers such Wollemi pine are a living link to the earth's first forests and the evolution of the seed habit over 300 million years ago. At present, the question centers on the phylogenetic relationships among Gnetales, other gymnosperms and angiosperms. This is controversial; others make the case for using evolution-driven taxonomy for drawing conclusions, rather than phylogeny alone (Farjon 2007). Answering the question of seed plant evolution does require all available information: DNA-based phylogeny, genomics, taxonomy and the fossil record (Brenner and Stevenson 2006) but the search for the common seed plant ancestor to both angiosperms and gymnosperms falls outside the scope of this book.

1.1 Gymnosperms

Gymnosperms, like all seed plants, have a diplohaplontic life cycle which alternates between a dominant sporophyte phase and a short gametophyte phase. Another feature is heterospory: seed plants have two kinds of spores, female and male, rather than a single bisexual spore. Heterospory and the seed habit set conifer reproductive biology apart from that of other green land plants such as mosses, liverworts and ferns.

Angiosperms are distinct from gymnosperms because they have an ovule enclosed in a carpel and gymnosperms lack this covering. Note that gymnosperms are defined by their *absence* of reproductive characters rather than by a shared set of characters. This absence-based grouping translates into a surprising degree of variability for reproductive morphology among conifers (Hart 1987).

1.2 Conifer Families: Classification and Geographic Distribution

While Wollemi pine is a conifer, it is not truly a pine. It belongs to the Araucariaceae, a Southern Hemisphere conifer family. The Order Coniferales refers to "cone-bearing plants" and the order includes seven families and roughly 620+ species. The seven conifer families are Pinaceae, Cupressaceae *sensu lato*, Araucariaceae, Podocarpaceae, Cephalotaxaceae, Taxaceae and most recently, the monotypic family Sciadopityaceae (Table 1.1). The history of taxonomic classification for the Coniferales is summarized in Stefanovic et al. (1998) who also showed DNA-based phylogenetic support for the seven conifer families as a monophyletic group.

Recent changes to taxonomy, particularly the merging of Cupressaceae and Taxodiaceae, have been well-described (Eckenwalder 1976; Price and Lowenstein 1989; Brunsfeld et al. 1994; Farjon 1998). The seven conifer families as a group have worldwide distribution (Li 1953) although Mexico, China, southeast Asia and New Caledonia are particularly rich centers of conifer species diversity.

Table 1.1 The distribution of the seven conifer families within the Coniferales. Wollemi pine belongs in the Araucariaceae

Family	Distribution
Araucariaceae	Tropical and southern temperate Southern Hemisphere
Cephalotaxaceae	Temperate and subtropical, east Asia
Cupressaceae *sensu lato*[a]	Mainly temperate, bihemispheric: east Asia, North America, New Caledonia
Pinaceae	Mostly Northern Hemisphere, a few tropical
Podocarpaceae[b]	Pantropical and temperate Southern Hemisphere
Sciadopityaceae	Japan
Taxaceae	Bi-hemispheric: eastern Asia, North America and Tasmania

[a]Includes the Taxodiaceae which added 10 genera and 16 species.
[b]Includes the Phyllocladaceae which formerly had one genus and five species.

The Pinaceae, as the largest of the seven families (Table 1.1), serves as the focus of this book. It includes only 11 genera, all of which are monoecious, yet comprises a total of 215 species of evergreen and deciduous needle-leaved conifers. Half of this Northern Hemisphere family belongs to a single genus, *Pinus*.

The genus *Pinus* is divided into two subgenera, the haploxylon or soft pines in subgenus *Strobus* and the diploxylon or hard pines in subgenus *Pinus*. The Pinaceae is considered a Northern Hemisphere family although there are tropical exceptions such as *Pinus caribaea* and *Pinus merkusii*. Many of the 100+ pine species have extensive ranges (*Pinus taeda*, *Pinus sylvestris*, *Pinus resinosa*) and others are even aggressive colonizers when planted as exotics (*Pinus radiata*; Richardson et al. 1994).

Other conifers are included in the Cupressaceae *sensu lato*, Sciadopityaceae, the two Southern Hemisphere families and the two taxa families (Table 1.2). For the Cupressaceae, most species are monoecious although a few are dioecious. Classifying this family's extant and fossil species is a dynamic and ongoing effort (Escapa et al. 2008). Consider *Xanthocyparis vietnamensis*, a new endemic species discovered in Vietnam (Farjon et al. 2002). New extant species are still being discovered and others are being re-classified as study specimens become available.

Members of a single genus, *Juniperus*, compose half of this family. Adding the Taxodiaceae family expanded the Cupressaceae by nine genera (Table 1.2) but several of these genera are monospecific such as *Sequoia*, *Metasequoia* and *Sequoiadendron* (Table 1.2). They have disjunct distributions which suggests that they are relics of groups once more abundant. In fact, *Metasequoia glyptostroboides* was once heralded as the discovery of a living fossil over 50 years ago is likely to have been one of several species in this genus (Merrill 1948).

The Southern Hemisphere conifer families are the Araucariaceae and the Podocarpaceae. These are the least-studied taxa but perhaps the richest source of biological exceptions to *Pinus* as a "type conifer." From chromosome number to unusual reproductive development, these exceptional Southern Hemisphere conifers should not even be assumed to grow as tall trees. A few of the Podocarpaceae are dwarf or diminutive perennials: *Microstrobos fitzgeraldii*, *Falcatifolium angustum* and *Lepidothamnus fonkii* (2008 IUCN Red List, www.redlist.org accessed January 19, 2009). The two families also provide a number of interesting reproductive exceptions although there are few genera or species in this group.

Box 1.1 Threatened and endangered conifers today

Massive extinction of many conifer and gymnosperm lineages came at the end of the Mesozoic era. For its few survivors, subsequent displacement, adaptation and speciation have shaped modern taxonomic distribution. How well do modern conifers fare? Many have large census population sizes, others are vulnerable but only 18 species are classified as critically endangered by the 2008 IUCN Red List for Threatened Species (www.iucnredlist.org, accessed on January 19, 2009). Wollemi pine is listed here because it has a census population of 50 trees.

Table 1.2 The 71 genera within the seven families of the Coniferales.[a] As a Southern Hemisphere conifer, Wollemi pine is closely related to *Agathis* spp

Family	Genera
Cupressaceae *sensu lato*[b]	*Actinostrobus, Austrocedrus, Callitris, Calocedrus, Chamaecyparis, X Cupressocyparis, Cupressus, Diselma, Fitzroya, Fokienia, Juniperus, Libocedrus, Microbiota, Neocallitropsis, Papuacedrus, Platycladus, Pilgerodendron, Tetraclinis, Thuja, Thujopsis, Widdringtonia, Xanthocyparis*
	Arthrotaxis, Cryptomeria, Cunninghamia, Glyptostrobus, Metasequoia, Sequoia, Sequoiadendron, Taiwania, Taxodium
Pinaceae	*Abies, Cathaya, Cedrus, Keteleeria, Larix, Nothotsuga, Picea, Pinus, Pseudolarix, Pseudotsuga* and *Tsuga*
Araucariaceae	*Araucaria, Agathis, Wollemia*
Podocarpaceae	*Acmopyle, Afrocarpus, Dacrycarpus, Dacrydium, Falcatifolium, Halocarpus, Lagarostrobos, Lepidothamnus, Manoao, Microcachrys, Microstrobos, Nageia, Parasitaxus, Phyllocladus, Podocarpus, Prumnopitys, Retrophyllum, Saxegothaea, Sundacarpus*
Sciadopityaceae	*Sciadopitys*
Cephalotaxaceae	*Cephalotaxus*
Taxaceae	*Amentotaxus, Austrotaxus, Pseudotaxus, Taxus, Torreya*

[a]Many genus names continue to be changed so the following chapters adhere the same genus name used by the original author; taxonomic conversions are required.

[b]The first group formerly belonged to Cupressaceae and the second group formerly belonged to Taxodiaceae; togther these make up the combined family of Cupressaceae *sensu lato*.

Table 1.3 A list of critically endangered conifer species worldwide. Species for only one of several higher-risk categories are shown here. Critically endangered refers to extremely high risk of extinction in the wild in the immediate future (2008 IUCN Red List of Threatened Species, accessed January 19, 2009)

Species	Range
Abies beshanzuensis	China: Zheijiang
Abies nebrodensis	Italy
Abies yuanbaoshanensis	China
Abies ziyuansis	China
Acmopyle sahniana	Fiji
Amentotaxus formosana	Taiwan
Araucaria angustifolia	Argentina, Brazil, Paraguay
Araucaria nemorosa	New Caledonia
Dacrydium guillauminii	New Caledonia
Juniperus bermudiana	Bermuda
Metasequoia glyptostroboides	China
Picea martinezii	Mexico
Podocarpus beecherae	New Caledonia
Podocarpus palawanensis	Philippines
Podocarpus perrieri	Madagascar
Taxus floridana	USA
Thuja sutchuenensis	China
Wollemia nobilis	Australia

The IUCN's Conifer Specialist Group lists eight of the 18 species as having reproductive problems as the cause for decline (Table 1.3). Among these eight species is *Wollemia nobilis* (Araucariaceae) in addition to *Abies beshanzuensis*, *Abies nebrodensis*, *Abies yuanbaoshanesis*, *Abies ziyuanensis* (Pinaceae); *Araucaria angustifolia* (Araucariaceae); *Thuja sutchenuensis* (Cupressaceae) and *Acmopyle sahniana* (Podocarpaceae). Exotic insect and fungal disease introductions, tourist development and overlogging of highly prized wood by local communities also contribute.

Other conifers, in addition to these 18 taxa, are identified on the IUCN Red List as endangered or vulnerable to extinction. Many at-risk taxa are indigenous to southwestern China or New Caledonia because these are centers of conifer species diversity (Contreras-Medina and Vega 2002). Southwestern China has a total of 31 genera and the highest diversity of gymnosperm species worldwide. Similarly, New Caledonia has a concentration of 17 genera and the highest concentration of Araucariaceae worldwide (Contreras-Medina and Vega 2002).

1.3 Fossil Record for Early Seed Plants

The terrestrial invasion of free-sporing land plants was followed by the advent of heterospory and with it, coordinated wind-pollination systems. Next came the seed habit. Cordaites, transition conifers and conifers had strobili. Reproductive features of modern conifers had evolved by the end of the Mesozoic era (Table 1.4).

Table 1.4 Major events for early land plants and gymnosperms on a geological time scale (Stewart and Rothwell 1993)

Era	Period	Beginning millions years (MY)	Climate	Plant life
	Cretaceous	141	No latitudinal temperature gradient; average temperatures higher than present-day; similar temperatures from poles to equator	Dinosaur extinction (65 MY)
				Spread of angiosperms
				First placental mammals
	Jurassic	195	Humid, warmer	Abundant dinosaurs, conifers, ginkgos, ferns, cycads, some pteridosperms, first birds
Mesozoic	Triassic	225	Arid, savanna type climate	Diversification of conifers
				Early mammals
				Rise of dinosaurs
				Pangaea begins to break apart into Laurasia and Gondwana
Paleozoic	Permian	280	Dry, seasonal fluctuations	Transition conifers, gingko, Cordaites
	Carboniferous	345	Late: polar glaciation with warm humid tropics	Cordaites, transition conifers, horsetails, mosses, ferns, seed ferns
			Early: Uniformly warm, humid climate	
	Upper			Pteridosperms
	Devonian		South Pole glaciation begins; warm, even climate	Lyginopterids First seed plants with hydrasperman reproduction
				Progymnosperms (*Archaeopteris*)
	Middle Devonian		Mild climate	Heterospory
	Early Devonian	395	Mild climate, heavy rainfall, aridity	Free-sporing land plants
Paleozoic	Silurian	435	Mild climate	First vascular plants
				Terrestrial invasion

1.3.1 Terrestrial Invasion of Land Plants

Land plants evolved from aquatic ancestors. The movement of plants to land, or the so-called terrestrial invasion, began at the end of the Silurian period during the Paleozoic era, roughly 435 MY (Table 1.4; Stewart and Rothwell 1993). The terrestrial invasion required adaptation to air rather than to land because early land plants had to adapt to the desiccating, aerial environment. In particular, terrestrial invasion required a drastic change in reproduction. Until now, reproduction had been dependent on the watery movement of gametes but now depended on aerial movement of spores (Chapman 1995).

Aerial movement of spores was one of many integral steps towards seed plant evolution. This adaptation enabled the colonization of land areas farther from water and this began the first greening of the earth's surface (Table 1.4). The next critical step was heterospory.

1.3.2 Heterospory

All free-sporing land plants prior to the Middle Devonian were dispersing a single bisexual spore (homospory) but by the Upper Devonian, a few plants had developed large female spores (700–900 μm) and small male spores (33–48 μm). This was the start of heterospory (Stewart and Rothwell 1993, p. 281) which coincided with the first Paleozoic forests, composed of *Archaeopteris*. Some species of *Archaeopteris* from the Upper Devonian also provided conclusive evidence of heterospory (Gensel and Andrews 1984; Kerp et al. 1990; Stewart and Rothwell 1993, pp. 263–278).

Archaeopteris was a progymnosperm, not a seed plant (Pettit and Beck 1968) so its reproduction was fern-like or pteridophytic. These plants had secondary cambium similar to gymnosperms (Stewart and Rothwell 1993, pp. 263–278). With the advent of heterospory, female and male sporogenesis diverged in form and function.

1.3.3 The Concept of Hydrasperman Reproduction

The hydrasperman pollination concept is constructed from an aggregate of reproductive characters found among many different early seed plants. These plants include series of fossil fern-like Paleozoic seed plants or pteridosperms including extinct *Archaeosperma* and *Hydrasperma* which appear in the fossil record from Late Devonian to the Permian (Andrews 1963; Gensel and Andrews 1984). The hydrasperman mating system is thought to have been wind-pollinated although a few insect-pollinated plants were present at this time (Taylor and Millay 1979; Matten et al. 1984; Rothwell and Scheckler 1988; Labandeira et al. 2007).

Hydrasperman plants lacked female strobili. Instead, naked preovules attached directly to a branch, subtended by wavy cupules. Paleozoic prepollen was large yet windborne. Some of these small plants more closely resembled primitive seed ferns or pteridosperms even though they had the seed habit (Stewart and Rothwell 1993, p. 300).

Radical adaptive radiation in these and other reproductive characters coincided with the advent of the first green land plant ecosystems (Rothwell and Scheckler 1988; Niklas 1997). Unimpeded colonization of terrestrial areas far from water opened new selection space and this spurred a high degree of adaptive radiation in hydrasperman reproduction (Niklas 1997). The selection pressure on the adult sporophyte would been greater for its pollination success than protection against herbivory, abiotic stress or even seed dispersal (Haig and Westoby 1989).

1.3.4 Cordaites: Tree-Like Gymnosperms

By the Carboniferous period, the Cordaites had appeared. Cordaites were tree-like gymnosperms which first appeared in the fossil record in the middle of the Early Carboniferous and persisted into the Permian (Costanza 1985; Stewart and Rothwell 1993, p. 410, Wang 1998; Looy et al. 2001). The cordaitean plants had female strobili enclosing megasporangia, in contrast to the hydrasperman plants which had exposed ovules attached to branches or leaves. Cordaites shared the callistophytalean reproductive habit in which both the micropyle and the nucellar beak within the pollen chamber were sealed after pollination. Male strobili of Cordaites were morphologically similar to conifer strobili (Mapes and Rothwell 1998).

1.3.5 Walchian Gymnosperms as Transition Conifers

The Walchian gymnosperms were a paraphyletic assemblage of now-extinct Late Carboniferous and Permian plants, considered to be transitional conifers (Stewart and Rothwell 1993, p. 410). These gymnosperms were tree-like with secondary wood and deep roots (Costanza 1985). They composed a dominant part of the forested landscape from the Permian to Triassic periods (DiMichele et al. 2001).

Pollen assemblages from the end of the early Carboniferous which are dominated by striated, bisaccate pollen which may derive from early transitional conifer forests, although little is known about the parent plants of these pollen assemblages. Walchian reproductive morphology is proposed as heterosporous and monoecious with separate micro- and megasporangia borne along fertile scales (Hernandez-Castillo et al. 2001). Body fossils of transitional conifers appear in the Late Carboniferous, overlapping with the cordaites. The last cordaites disappeared from the stratigraphic record at the Permian-Triassic boundary (Looy et al. 2001). By the late Triassic, conifers and other gymnosperms had become the dominant forest type ranging from tropical to boreal latitudes (Looy et al. 2001).

1.3.6 Rapid Conifer Diversification in the Mesozoic Era

At the end of the Triassic period, the earth had become drier and warmer. New mountain ranges now served as physical and climatic barriers which slowed dispersal and colonization of fauna and flora. The major plant community changed from lycopods to seed plants and this change was complete by the Triassic period.

By the late Triassic, the dinosaurs appeared and conifers were prevalent. Complex ecosystems began to evolve within these forests as evidenced by the commensal symbioses among dung beetles, conifers and herbivorous dinosaurs (Chin and Gill 1996).

This was followed by a period of rapid conifer diversification, the Golden Age of Conifers. Climatic conditions favored rapid diversification of conifers, gingkos and other gymnosperms. The ecological dominance of conifers was at its zenith. Far more conifer lineages were present than exist today (Niklas et al. 1983; Knoll 1986; Fig. 1.1).

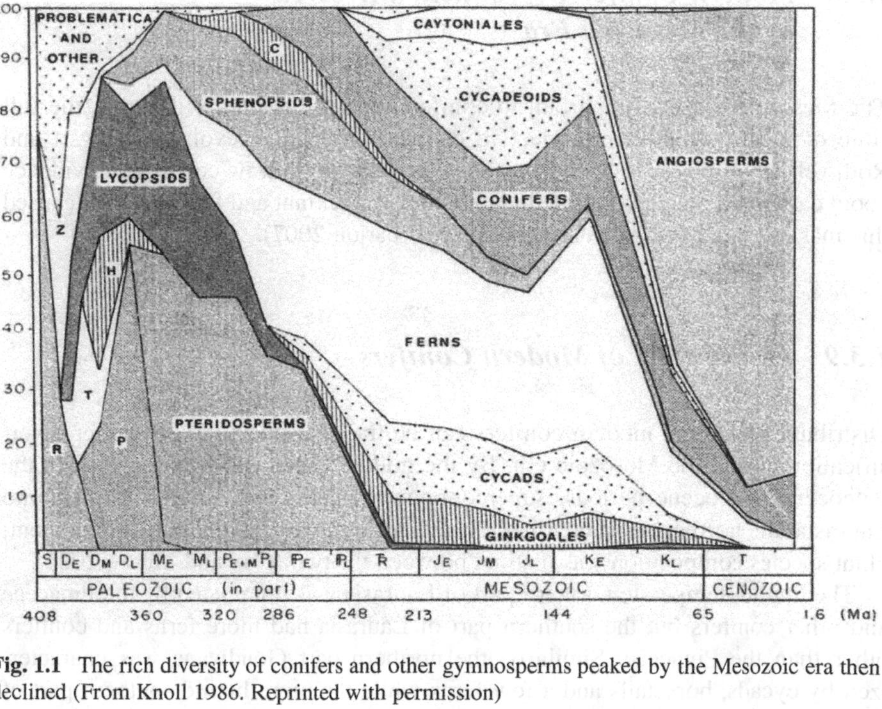

Fig. 1.1 The rich diversity of conifers and other gymnosperms peaked by the Mesozoic era then declined (From Knoll 1986. Reprinted with permission)

1.3.7 Rise and Spread of Angiosperms

Angiosperms first appear in the early Cretaceous, roughly 135 MY. Many modern Northern Hemisphere angiosperm families were recognizable by 95 MY ago so differentiation must have been rapid. Angiosperms diversified and became dominant at low latitudes up to 20° from the equator. They composed 60–80% of low-latitude floras but only 30–50% of the high-latitude floras.

Rapid spread of angiosperm plants brought the Golden Age of conifers to an end. The number of conifer species dropped sharply. Surviving conifer lineages were displaced to more extreme environments at either higher elevations or higher latitudes where limited diversification continued, producing present-day conifer species (LePage 2003).

Note that conifers and other gymnosperm lineages had already survived cataclysmic events so harsh that marine life and later dinosaurs were extirpated. This suggests that the equatorial spread of angiosperm plants was formidable. Many conifers and gymnosperm lineages eventually met with extinction (Fig. 1.1). Only a fraction survived; these were displaced to higher elevations or higher latitudes (Looy et al. 1999). Modern conifers and gymnosperms thus represent a small fraction of what once existed of a rich Mesozoic flora (Knoll 1986; Fig. 1.1).

1.3.8 Modern Conifer Reproduction Evolved by the Mesozoic Era

The Mesozoic era was important to conifers for another reason: by now, the full suite of modern conifer reproductive characters had now evolved (Stewart and Rothwell 1993, pp. 425–429). Even so, it is difficult to draw conclusive evidence about early seed plant evolution due to the missing extant and fossil records caused this massive late Mesozoic extinction event (Farjon 2007).

1.3.9 Distribution of Modern Conifers

Distribution of some modern conifers can be traced to the peak of conifer diversification during the Mesozoic era. By the middle Cretaceous periods, before the Paleocene or Eocene periods, supercontinent Pangaea had already divided into Laurasia, the northernmost continent, and Gondwana, the southernmost continent; plant species composition was distinct between the two (Cox and Moore 2005).

The humid, temperate northern part of Laurasia was dominated by the Pinaceae and other conifers but the southern part of Laurasia had more ferns and conifers other than the Pinaceae. Similarly, the northern part Gondwana was characterized by cycads, horsetails and a few ferns while the more humid southern part of

Gondwana had many podocarps, araucarian conifers and ferns (Cox and Moore 2005). Modern conifer distributions still adhere to this pattern (Li 1953); as an example, *Wollemia nobilis* is indigenous to New South Wales, Australia which was once part of southern Gondwana.

1.4 Fossil Record for the Pinaceae

The fossil record suggests that the Cupressaceae family is older than the Pinaceae, having evolved by the Late Triassic. Its modern genera were already in evidence before the end of the Mesozoic (Miller 1977).

By contrast, the earliest fossil record for the Pinaceae extends as least as far as the Late Cretaceous and possibly to the late Jurassic (Miller 1977). For a long period, the oldest evidence was thought to come from a single cone, *Pinus belgica* (Alvin 1960) but this no longer holds true. *Pseudolarix* is now known to be the oldest representative of the Pinacaeae in the fossil record (Keller and Hendrix 1997; LePage 2003) and this has been confirmed using DNA-based phylogenetic analyses (Gernandt et al. 2008).

The *Pseudolarix* fossils were found in the Upper Jurassic Tsagaan Tsav Formation in southeastern Mongolia (43°34′54″ N, 108°06′12″ E) and they were dated at 156 MY (Table 1.5; Keller and Hendrix 1997; LePage 2003). Two *Pseudolarix* fossil species have been identified based on characteristics of the seed cone scale (LePage and Basinger 1995): they are similar to extant species *Pseudolarix amabilis* and to the fossil *Pseudolarix wehrii*, a rare and extinct species found only in North America at three Eocene-age locations (LePage and Basinger 1995). Like Wollemi pine, *Pseudolarix ambilis* is yet another living fossil dating back to the Jurassic period.

The genus *Pinus* has no known place or time although this has been the subject of considerable conjecture. Similarly, no fossil evidence has yet been found to estimate the time period before the genus split into haploxylon (*Strobus*) and diploxylon (*Pinus*) subgenera (Miller 1977). The oldest *Pinus* spp. fossil is a Lower Cretaceous lignitic seed cone, *Pinus belgica* (Alvin 1960) found in Belgium's Wealden Formation (ca. 140 MY). The fossil cone resembles a hard pine from *Contortae*, *Oocarpae* or *Sylvestres* subsections (Alvin 1960; Miller 1977). From this same time period, the subgenus *Strobus* is represented by a single cone from the Magothy Formation in Delaware (Miller 1977). *Pinus magothensis* Penny is considered similar to cones in the fossil form genus *Pityostrobus* (Miller 1977). However, note that these incomplete fossil records and time interval estimations have been re-estimated via molecular calibration (Willyard et al. 2007).

Genera in the Pinaceae other than *Pseudolarix* and *Pinus* may have evolved by the Cretaceous (Miller 1977) or perhaps the Upper Cretaceous and early Tertiary (Stewart and Rothwell 1993, pp. 425–429; LePage 2003). These genera are thought to have descended from an ancestral complex consisting of several different species of *Pityostrobus* (LePage 2003).

Table 1.5 Major events for the divergence and adaptation of the Pinaceae and the genus *Pinus*

Era	Period	Epoch	Beginning millions years (MY)	Climate	Plant life
Cenozoic		Holocene	10,000		Present-day
	Quaternary	Pleistocene	2.5	Polar ice caps form; glaciation	Major range shifts for existing species
		Pliocene	7		*Australes* (North American) subsect. of hard pines 11–12 MY
		Miocene	26	Continental uplifts	*Oocarpae* (North America) subsect. of hard pines 21–22 MY
		Oligocene	38	Ice on South Pole, not North Pole; warm temperate forests	
		Eocene	54	Subtropical climates, heavy rainfall, volcanism and mountain-building	Arctic *Metasequoia*-dominated conifer forests
	Tertiary	Paleocene	65	Warmer climate than present-day; subtropical	
Mesozoic	Cretaceous		141	Stable, equable climate from poles to equator	Diversification of the Pinaceae; fossils *Pinus belgica* and Magothy Formation
					Dinosaur extinction (65 MY); spread of angiosperms
	Jurassic		195	Humid, warmer	First Pinaceae fossil is *Pseudolarix* from Mongolia
					Araucaroid fossil species here is similar to Wollemi pine: *Agathis jurassica*
					Peak of the dinosaur era
Mesozoic	Triassic		225	Arid, savanna type climate	Diversification of conifers
					First dinosaurs
					Pangaea breaks into Laurasia and Gondwana

Another influential factor towards the persistence of the Pinaceae is its symbiotic ectomycorrhizal associations. This plant–fungi association is thought to have contributed to the rapid adaptation of pines to extreme environments after the diaspora (LePage et al. 1997). Evidence comes from the earliest known ectomycorrhizal fossils on *Pinus* spp. roots in the Middle Eocene, an otherwise rare symbiont among flowering plants (LePage et al. 1997). Similarly, the Pinaceae (Brundett 2002) are the predominant gymnosperm taxon with ectomycorrhyizae; most Southern Hemisphere gymnosperm genera such as *Podocarpus*, *Araucaria* and *Agathis* have only vescicular-arbuscular mycorrhizae (Brundett 2002).

Pines flourished in abundance in North America, Asia and Europe during this time period. Many modern-day lineages can be traced to Miocene ancestors (Table 1.5; Millar 1993; Krupkin et al. 1996). In North America, hard pine subsection *Contortae* is thought to be among the oldest hard pine lineages North America with divergence dating back to 76–86 MY (Table 1.5; Krupkin et al. 1996) but one cannot dismiss the possibility that more ancient hard pine lineages once existed or have yet to be found (Farjon 1996). Molecular evidence for these shallow speciation events is highly controversial and yet to be fully resolved. Molecular data suggests that the subsection *Oocarpae* is more recent, thought to have evolved around 34–64 MY. The progenitor of hard pine subsection *Australes* is hypothesized to have arisen from an *Oocarpae* ancestor 10–12 MY, migrating from west to east during the Miocene (Krupkin et al. 1996). If so, the subsection *Australes*, which includes *Pinus taeda*, is one of the younger hard pine clades in North America (Krupkin et al. 1996), as shown in Table 1.5.

1.5 Examples of Other Conifers as Living Fossils

Wollemia nobilis is a living fossil, similar to *Pseudolarix amabilis* and the two pine subgenera, *Pinus* and *Strobus*. But two more conifer species described below also illustrate this point.

1.5.1 The Metasequoia-*Dominated Forest in Canadian High Arctic*

Metasequoia glyptostroboides is well-known as a relic species found in remote areas of China (Merrill 1948). Its offspring grow in botanical gardens worldwide but it is remains on the IUCN's critically endangered list because so its original census populations are so small and fragmented (Table 1.3). Contrast this fragile species with the vast *Metasequoia*-dominated conifer forests that once grew in the Canadian High Arctic (Photo 1.1).

Starting in the Mesozoic era, these conifer forests grew in Asia then extended into high latitudes above 70° (LePage 2003; Cox and Moore 2005) in places such

Photo 1.1 Abundant fossil remains of a conifer forest preserved at high latitudes middle Eocene deposits on Axel Heiberg Island, Nuavit Territory Canada. These *Metasequoia*-dominated forests grew north of the Arctic Circle at paleo-latitudes 75° to 80°C in the Canadian High Arctic (Photograph by Ben LePage. Permission granted)

as Axel Heiberg Island and Ellesmere Island in the Canadian Arctic (81° N) but they did not resemble any present-day forest (LePage 2003). These were high-biomass forests dominated by *Metasequoia* (Williams et al. 2003; Jahren 2007; LePage et al. 2005) but also inclusive of *Picea*, *Pinus*, *Keteleeria*, *Tsuga* (Pinaceae) and *Glyptostrobus*, *Taiwania*, *Thuja*, *Chamaecyparis* and *Cathaya* (Cupressaceae) (LePage 2003). A few temperate broad-leaved deciduous forest species were also present (LePage et al. 2003). Later, from the Middle Eocene to the Early Miocene, the earth's climates became cooler, drier and more seasonal until forests could no longer be sustained. Only fossilized remnants of this Arctic conifer forest can still be seen (Photo 1.1).

1.5.2 **Pinus krempfii** *in the Da Lat Plateau of Vietnam*

The odd-looking Krempf's pine (deFerré 1948) depicted in Photo 1.2 is regarded as a living pine fossil because its flat needle morphology and biogeography (Buchholz 1951). Its flat-leaved needles bear a striking similarity to fossil conifers. Even now,

Photo 1.2 *Pinus krempfii* has long been considered a living fossil due to its flat needles. It is now classified as a member of the genus Pinus, subgenus *Strobus*. The species is endemic to the Da Lat Plateau in the central highlands of Vietnam. Shown are the first-year conelet along with the flat needles

P. krempfii reproductive biology is poorly understood for a host of reasons: the few individuals left of this species grow only in a high-elevation mountain range, they shed seeds during monsoon season, seedlings do not thrive in arboreta outside the Da Lat Plateau and few herbaria outside of Vietnam have specimens.

Historical events kept the taxonomy of *Pinus krempfii* controversial until the latter twentieth century. In 1944, Chevalier elevated *Pinus krempfii* to a monotypic genus and renamed it *Ducampopinus krempfii*. Later, deFerré (1948) considered this to be premature, but the change was retained. This divided the genus *Pinus* into three (not two) subgenera: *Pinus*, *Strobus* and *Ducampopinus*. Later, the single vascular bundle in the needle and heartwood phenolics suggested *Pinus krempfii* belonged to subgenus *Strobus* (Stephan and Tien 1986). Further corroboration came with a molecular phylogeny which clearly places *Pinus krempfii* within the subgenus *Strobus* (Wang et al. 2000). The third subgenus, *Ducampopinus*, is no longer considered to be valid.

The other supporting evidence of *Pinus krempfii* as a living fossil comes from the biogeography of the Da Lat Plateau. The ancient endemic ecosystem in the Da Lat Plateau survived planetary climate change, rising oceans and glacial movement. The vegetation of the Da Lat Plateau was likely protected by a moderating oceanic influence unaffected by glaciation. Its temperate forest composition is typical of the Tertiary period in the Northern Hemisphere (Wen 1999).

A few *Pinus krempfii* populations have survived the defoliating herbicides, land-clearing and high-explosion munitions during war (Westing and Westing 1981; Stephan and Tien 1986). *Pinus krempfii* is now protected by *in situ* preserves and by *ex situ* conservation programs also located within the Da Lat Plateau. Three national *in situ* preserves have been established for *Pinus krempfii* and each now has a few hundred individuals at advanced ages. Preventing fires has resulted in a vigorous hardwood understory at the expense of *Pinus krempfii* seedling regeneration. Local foresters are successfully transplanting seed and seedlings to *ex situ* conservation banks within the Da Lat region. Today, the IUCN classifies *Pinus krempfii* as vulnerable rather than endangered but this is adequate for justifying its program in preservation and rehabilitation of Vietnamese central montane forests.

Another interesting feature of the Da Lat Plateau is that it is part of the eastern U.S.-Asia parallel (for review see Boufford and Spongeford 1983). This refers to the parallel floral composition in both places. Many of the same angiosperm plant species found here in the Da Lat Plateau forests are also indigenous to Atlantic Seaboard forests of the United States.

The Da Lat Plateau, much of Vietnam and southeast Asia is more richly endowed with conifer diversity by world standards. *Pinus krempfii* and five other pines species grow along a continuum from south-central to northern Viet Nam: another endemic, *P. dalatensis* de Ferré (subgenus *Strobus*, subsection *Strobi*) and four more widespread species: *Pinus kesiya* Royle (subgenus *Pinus*, subsection *Sylvestres*), *Pinus merkusii* De Vriese (subgenus *Pinus*, subsection Sylvestres), *Pinus massioniana* Lamb. (subgenus *Pinus*, subsection *Sylvestres*) and *Pinus yunnanensis* Franchlet (subgenus *Pinus*, subsection *Sylvestres*) (Mirov 1967). More than pines, other conifer genera are found in this part of southeast Asia (LePage 2003). Such rich conifer species diversity can also be found in mountain ranges in Mexico, New Caledonia, southwest China and to a lesser extent, North America, all of which serve as present-day centers of conifer species diversity (Mirov 1967; Millar 1993; Farjon 1996; Farjon and Styles 1997).

1.6 Closing

In retrospect, Wollemi pine is neither a pine nor even a member of the Pinaceae: it is a member of the Araucariaeae, a Southern Hemisphere conifer family. But finding Wollemi pine in New South Wales was a remarkable discovery because it resembles the fossil of another araucaroid conifer, *Agathis jurassica*, and such Jurassic conifer fossils are rare. The discovery of Wollemi pine also renews public interest in conifers as Mesozoic relics.

Wollemia nobilis, *Pseudolarix amabilis*, *Metasequoia glyptostroboides* and *Pinus krempfii* – these and other conifers are living fossils. In a broad sense, all modern conifers are living fossils because they are Mesozoic relics. With the equatorial spread of angiosperms, conifers were displaced to high-latitude or high-elevation environments. Evidence of this displacement can be seen from the remnants of the *Metasequoia*-conifer forests in the Arctic Circle and from the endemic *Pinus krempfii* in Vietnam. What are features of conifers might confer resilience to rapid change?

References

Alvin, K. 1960. Further conifers of the Pinaceae from the Wealden Formation of Belgium. Institut Royal des Sciences Naturelles de Belgique Memoires 146: 1–39.

Andrews, H. 1963. Early seed plants. Science 142: 925–931.

Boufford, D. and S. Spongeford. 1983. Eastern Asia-eastern North American phytogeographical relationships – a history from the time of Linnaeus to the twentieth century. Annals of the Missouri Botanical Garden 70: 423–439.

Brenner, E. and D. Stevenson. 2006. Using genomics to study evolutionary origins of seeds. Editor: C.G. Williams. In: *Landscapes, Genomics and Transgenic Conifers*. Springer, Dordrecht, The Netherlands. pp. 85–106.

Brundett, M. 2002. Coevolution of roots and mycorrhizas of land plants. New Phytologist 154: 275–304.

Brunsfeld, S., P. Soltis, et al. 1994. Phylogenetic relationships among the genera of Taxodiaceae and Cupressaceae. Systematic Botany 19: 253–262.

Buchholz, J. 1951. A flat-leaved pine from Annam, Indo-china. American Journal of Botany 38: 245–252.

Chapman, D. 1995. Plant transitions to land Chapter 3. Editors: M. Gordon and E. Olson. In: *Invasions of the Land: The Transition of Organisms from Aquatic to Terrestial Life*. Columbia University Press, New York. 312 pp.

Chin, K. and B. Gill. 1996. Dinosaurs, dung beetles and conifers: participants in a Cretaceous food web. Palaios 11: 280–285.

Contreras-Medina, R. and I. Vega. 2002. On the distribution of gymnosperm genera, their areas of endemism and cladistic biogeography. Australian Systematic Botany 15: 193–203.

Costanza, S. 1985. *Pennsylvanioxylon* of middle and upper Pennsylvanian coals from the Illinois basin and its comparison with *Mesoxylon*. Palaeontographica Abt. B 197: 81–121.

Cox, C. and P. Moore. 2005. *Biogeography: An Ecological and Evolutionary Approach*. Blackwell, Malden, MA, Seventh Edition, 440 pp.

deFerré, Y. 1948. Quelques particularities anatomiques d'un pin indochinois: *Pinus krempfii*. Bulletin de la société d'histoire naturelle de Toulouse 83: 1–6.

DiMichele, W., S. Mamay, et al. 2001. An early Permian flora with late Permian and Mesozoic affinities for north-central Texas. Journal of Palaeontology 75: 449–460.

Eckenwalder, J. 1976. Re-evaluation of Cupressaceae and Taxodiaceae: a proposed merger. Madrono 23: 237–256.

Escapa, I., R. Cuneo, et al. 2008. A new genus of the Cupressaceae (*sensu lato*) from the Jurassic of Patagonia: implications for conifer megasporangiate cone homologies. Review of Palaeobotany and Palynology 151: 110–122.

Farjon, A. 1996. Biodiversity of *Pinus* (Pinaceae) in Mexico: speciation and paleo-endemism. Botanical Journal of the Linnaean Society 121: 365–384.

Farjon, A. 1998. *World Checklist and Bibliography of Conifers*. Royal Botanic Gardens, Kew.

Farjon, A. 2007. In defence of a conifer taxonomy which recognizes evolution. Taxon 56: 639–641.

Farjon, A. and B. Styles. 1997. *Pinus* (Pinaceae). New York Botanical Garden, New York.

Farjon, A., T. Nguyen, et al. 2002. A new genus and species in Cupressaceae (Coniferales) from Northern Vietnam, *Xanthocyparis vietnamensis*. Novon 12: 179–189.

Gensel, P. and H. Andrews. 1984. *Plant Life in the Devonian*. Praeger, New York.

Gernandt, D., S. Magallan, et al. 2008. Use of simultaneous analyses to guide fossil-based calibrations of Pinaceae phylogeny. International Journal of Plant Sciences 169: 1086–1099.

Haig, D. and M. Westoby. 1989. Selective forces in the emergence of the seed habit. Biological Journal of the Linnean Society 38: 215–238.

Hart, J. 1987. A cladistic analysis of conifers: preliminary results. Journal of Arnold Arboretum 68: 269–307.

Hernandez-Castillo, G., G. Rothwell, et al. 2001. Thucydiaceae Fam. Nov., with a review and re-evaluation of Paleozoic Walchian conifers. International Journal of Plant Sciences 162: 1155–1185.

Jahren, A. 2007. The Arctic forest of the middle Eocene. Annual Review of Earth and Planetary Sciences 35: 509–540.

Keller, A. and M. Hendrix. 1997. Paleoclimatologic analysis of a Late Jurassic petrified forest, south-eastern Mongolia. Palaios 12: 282–291.

Kerp, J., R. Poort, et al. 1990. Aspects of Permian paleobotany and palynology: conifer dominated rotliegend floras from the Saar-Nahe Basin (Late Carboniferous Early Permian, SW Germany) with special reference to the reproductive biology of early conifers. Review of Paleobotany and Palynology 62: 205–248.

Knoll, A. 1986. Patterns of change in plant communities through geological time. Editors: J.A. Diamond and T.J. Case. In: *Community Ecology*. Harper & Row, New York. Chapter 7, pp. 126–141.

Krupkin, A., A. Liston, et al. 1996. Phylogenetic analysis of the hard pines (*Pinus* subgenus *Pinus*, Pinaceae) from chloroplast DNA restriction site analysis. American Journal of Botany 83: 489–498.

Labandeira, C., J. Kvacek, et al. 2007. Pollination drops, pollen and insect pollination of Mesozoic gymnosperms. Taxon 56: 663–695.

LePage, B. 2003. The evolution, biogeography and paleoecology of the Pinaceae based on fossil and extant representatives. Acta Horticulturae 615: 29–52.

LePage, B. and J. Basinger. 1995. Evolutionary history of the genus *Pseudolarix* Gordon (Pinaceae). International Journal of Plant Sciences 156: 910–950.

LePage, B., R. Currah, et al. 1997. Fossil ectomycorrhizae from the Middle Eocene. American Journal of Botany 84: 410–412.

LePage, B., B. Beauchamp, et al. 2003. Late early Permian plant fossils from the Canadian High Arctic: a rare paleoenvironmental/climatic window in northwest Pangea. Palaeogeography, Palaeoclimatology, Palaeoecology 191: 345–372.

LePage, B., C. Williams, et al. 2005. *The Geobiology and Ecology of Metasequoia*. Springer, Dordrecht, The Netherlands.

Li, H. 1953. Present distribution and habits of the conifers and the taxads. Evolution 7: 245–261.

Looy, C., W. Brugman, et al. 1999. The delayed resurgence of forests after the Permian-Triassic ecologic crisis. Proceedings National Academy of Sciences U.S.A. 96: 13857–13862.

Looy, C., R. Twitchett, et al. 2001. Life at the end: Permian dead zone. Proceedings National Academy of Sciences U.S.A 98: 7879–7883.

Mapes, G. and G. Rothwell. 1998. Primitive pollen cone structure in upper Pennylsvanian (Stephanian) Walchian conifers. Journal of Paeontology 72: 571–576.

Matten, L., T. Fine, et al. 1984. The megagametophyte of *Hydrasperma tenuis* long from the uppermost Devonian of Ireland. American Journal of Botany 71: 1461–1464.

Merrill, E. 1948. *Metasequoia*, another living fossil. Arnoldia 8: 1–8.

Millar, C. 1993. The impact of the Eocene on the evolution of *Pinus* L. Annals of the Missouri Botanical Garden 80: 471–498.

Miller, C. 1977. Mesozoic conifers. Botanical Gazette 43: 217–280.

Mirov, N. 1967. *The Genus Pinus*. Ronald Press, New York.

Niklas, K., B. Tiffney, et al. 1983. Patterns in vascular land plant diversification. Nature 303: 614–616.

Niklas, K. 1997. *The Evolutionary Biology of Plants*. University of Chicago Press, Chicago, IL.

Pettit, J. and C. Beck. 1968. *Archaeosperma arnoldii* - a cupulate seed from the Upper Devonian of North America. Contrib. Mus. Paleont. Univ. Michigan 22: 139–154.

Price, R. and J. Lowenstein. 1989. An immunological comparison of Sciadopityaceae, Taxodiaceae and Cupressaceae. Systematic Botany 14: 141–149.

Richardson, D., P. Williams, et al. 1994. Pine invasions in the Southern Hemisphere: determinants of spread and invadibility. Journal of Biogeography 21: 511–527.

Rothwell, G. and S. Scheckler. 1988. Biology of ancestral gymnosperms. Editor: C. Beck. In: *Origin and Evolution of Gymnosperms*. Columbia University Press, New York. pp. 85–134.

Stefanovic, S., M. Jager, et al. 1998. Phylogenetic relationships of conifers inferred from partial 28S rRNA gene sequences. American Journal of Botany 85: 688–697.

Stephan, G. and L. Tien. 1986. Development of pine resin production in Vietnam (translated from German). Soz. Forst. 36: 120–121.

Stewart, W. and G. Rothwell. 1993. Paleobotany and the evolution of plants. Cambridge University Press, New York. 521 p.

Taylor, T. and M. Millay. 1979. Pollination biology and reproduction of early seed plants. Review of Paleobotany and Palynology 27: 329–355.

Wang, S.-J. 1998. The cordaitean fossil plants from Cathaysian area in China. Acta Botanica Sinica 40: 573–579.

Wang, X.-R., A. Szmidt, et al. 2000. The phylogenetic position of the endemic flat-needle pine *Pinus krempfii* (Pinaceae) from Vietnam, based on PCR-RFLP analysis of the chloroplast DNA. Plant Systematics and Evolution 220: 21–36.

Wen, J. 1999. Evolution of eastern Asian and eastern North American disjunct distributions in flowering plants. Annual Review of Ecology and Systematics 30: 421–455.

Westing, A. and C. Westing. 1981. Endangered species and habitats of Vietnam. Environmental Conservation 8: 59–63.

Williams, C., A. Johnson, et al. 2003. Reconstruction of Tertiary *Metasequoia* forests. I. Test of a method for biomass determination based on stem dimensions. Paleobiology 29: 256–270.

Willyard, A., J. Syring, et al. 2007. Fossil calibration of molecular divergence in *Pinus*: inferences for ages and mutation rates. Molecular Biology Evolution 24: 90–101.

Chapter 2
The Diplohaplontic Life Cycle

Summary Seed plants have a diplohaplontic life cycle which has two phases. The first is a dominant, concurrent sporophyte phase and the second is a brief gametophyte phase; both phases are multicellular so somatic growth follows meiosis and somatic growth follows the union of gametes or syngamy. But even so, there are variants within this basic plan. Some angiosperm species, like bamboo, flower only once at the end of life. Conifers, by contrast, are long-lived iteroparous species which can produce male and female gametophytes annually for hundreds or even thousands of years. Each year, the adult sporophyte gives rise to a new set of indeterminate meristems at the telescoping ends of the latest branch tips and some of these branch tips develop either male or female reproductive initials. Conifers thus have meristems which produce both vegetative and reproductive growth as opposed to the predetermined germline typical of the diplontic life cycle of vertebrates. This "moving interface" between vegetative and reproductive cell lineages has not been well-studied in conifers and it constitutes a fertile area for exploring adaptive response, mutational accumulation, DNA repair systems and regulatory cues.

A surprising variety of life cycles is found among eukaryotes, especially for fungi, algae and some of the lower vascular plants (Coelho et al. 2007). The diplohaplontic life cycle of seed plants is often confusing, especially when compared to vertebrates, because the gametophyte phase has no direct analogy to the human condition. To see this, compare the three most common life cycles among eukaryotes as shown in Fig. 2.1: diploid, haploid and diplohaplontic life cycle types (Mable and Otto 1998).

(a) **Diplontic life cycle**. Life cycle typical of humans and other vertebrates. Meiosis is followed immediately by union of gametes or syngamy, not mitoses (see Fig. 2.1a). Somatic growth occurs only in the diploid stage. Single-celled gametes develop from predetermined germline tissues; this is the limited extent of the haploid phase.

(b) **Haplontic life cycle**. Life cycle is typical of green algae and some fungi. Syngamy is followed by meiosis (see Fig. 2.1b). Somatic growth occurs only in the haploid phase.

C.G. Williams, *Conifer Reproductive Biology*,

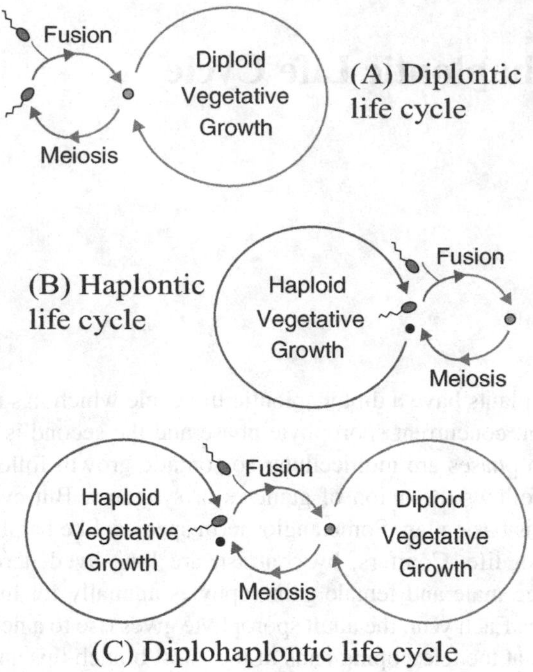

Fig. 2.1 Diplontic, haplontic and diplohaplontic life cycles are the three common life cycles for eukaryotes. They differ in the ordering of meiosis, syngamy and somatic growth (B. Mable and S. Otto. (1998) The evolution of life cycles with haploid and diploid phase. BioEssays 20: 453–462. Reprinted with permission from John Wiley & Sons, Inc.)

(c) Diplohaplontic life cycle. This is the life cycle is typical of many green plants: ferns, mosses as well as seed plants. Meiosis is followed by the somatic growth of the gametophyte but syngamy is also followed by the somatic growth of the diploid zygote (see Fig. 2.1c). The phases can be independent or dependent (i.e. one phase grows on the other).

In mosses, the gametophyte phase is dominant and larger than the sporophyte phase. The reverse is true for seed plants and some ferns where the diploid sporophyte phase is dominant and larger than the gametophyte phase. But in all green plants, the gametophyte is a multicellular product of mitotic divisions. These mitoses occur between meiosis and gamete formation such that gametes develop from the gametophyte, not the adult sporophyte.

The diplohaplontic life cycle was first recognized by Hofmeister in 1851 (cited in Mable and Otto 1998) but it was another 50 years before Strasburger (1894) discovered the alternating cycle of diploidy and haploidy, reporting on the periodic reduction of chromosomes during meiosis and subsequent union of gametes. In addition to these enduring discoveries, both Hofmeister and Strasburger made significant contributions to gymnosperm reproductive biology. These contributors and other eighteenth and nineteenth century botanists established a sturdy foundation of knowledge that continues to build (Singh 1978, pp. 4–5; Skinner 1992).

2.1 An Overview of the Conifer Life Cycle

The diplohaplontic life cycle of any seed plant is framed around five critical characters: (1) heterospory or separate spore types for male and female reproduction, (2) multicellular male and female gametophytes, (3) retention of the female gametophyte within the adult sporophyte, (4) the siphonogamous pollen tube and (5) dispersal of mature embryos. In addition, conifers share four other life cycle features which distinguish them from other seed plants:

Conifers have separate male and female strobili. Conifers are monosporangiate; each strobilus is either male or female but not both. This is an early consequence of heterospory, one of the five defining features of the seed plant life cycle. Separate male and female reproductive structures may form on a single tree (monecy or "one house") as in the case of all members of the Pinaceae family. But monoecious conifers are by no means hermaphroditic. Hemaphroditism is an aberrant state; as an example, a rare case was reported for a single *Pinus nigra* tree in Greece where a single strobilus had both male and female morphology (Matziris 2002). Many conifers have either male or female strobili on a single tree; this condition is dioecy ("two houses").

Diploid and haploid phases are concurrent. The adult sporophyte gives rise to the transient gametophyte phase annually; these two phases occur at the same time. This is apparent for the endosporic female gametophyte developing inside the ovule but it is less apparent for the male gametophyte. The male gametophyte is also dependent on the sporophyte for its protection from meiosis to fertilization. The phrase "alternating generations" is not apt in this case because diploid and haploid phases occur at the same time. Referring to diploid and haploid phases is more appropriate terminology for the diplohaplontic life cycle of conifers (Fig. 2.2).

Selfing occurs only as geitonogamy. Geitonogamy refers to the special case of selfing between separate male and female strobili on the same plant (Richards 1997). Conifers do self but they are only capable of geitonogamy, not autogamy. Autogamy refers to selfing within a single perfect flower where both male and female parts are present (Richards 1997).

Recurrent production of seeds (iteroparity). Once a conifer reaches reproductive onset, this plant is capable of producing male and female gametophytes each year for hundreds or even thousands of years.

Even so, the sporophyte has three stages: a juvenile stage, reproductive onset and reproductive competence. Juvenile is defined as the absence of any reproduction, even in the presence of strong stimuli. Next is the reproductive onset stage where reproduction is the exception rather than the rule: strong external stimuli are required before either male or female strobili will develop. After reproductive onset comes the reproductive competence stage where strobili develop annually under almost any conditions. Few authors make this delicate distinction so the historical literature can be confusing.

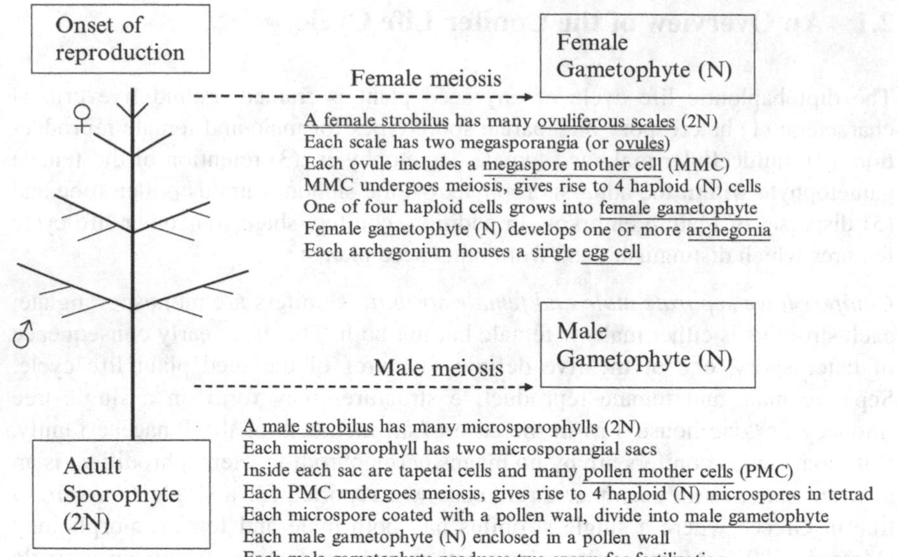

Fig. 2.2 Diplohaplontic life cycle for *Pinus taeda*

2.2 Juvenility

Characterized by the absence of reproduction, the juvenile period may also be marked by characteristic shoot morphology (Wareing 1959; Kozlowski 1971; Greenwood 1984). This period can lasts from 1 to 15 years for *Pinus taeda* (Dorman 1976). By contrast, juvenility can be as short as 2 years for *Pinus banksiana*, *Pinus virginiana* and several Cupressaceae taxa.

2.3 Reproductive Onset Versus Reproductive Competence

Reproductive onset occurs when strobili can be produced only in response to strong external stimuli. External stimuli include out-of-phase dormancy treatments (Greenwood 1978) or gibberellin applications, daylength changes or water stress (Burris et al. 1992; Bramlett et al. 1995). The effect of gibberellins on inducing male and female strobili in other conifers including the Cupressaceae is particularly well-documented (Pharis et al. 1987; Bonnet-Masimbert and Webber 1995). Reproductive onset is reported to begin at 5–10 years for *Pinus sylvestris* (Wareing 1959) or 10 years for *Pinus taeda* (Dorman 1976). Age of reproductive onset generally refers to either male strobili or female strobili because it is rare that both types of strobili appear at the same age. In many species, female strobili often emerge sooner than male strobili.

The third stage, reproductive competence, is defined as the earliest age at which either female or male strobili develop (Sax 1962) regardless of external stimuli. The age of reproductive onset is not equivalent to the age of reproductive competence (Kozlowski 1971).

The age of reproductive onset is also not a synonym for generation interval or generation length. Both generational terms are central concepts for population genetics theory and molecular evolution. Instead, generation interval is defined as the age when both male and female strobili appear on the majority of the trees in a given population; in practice, this population-level measure will be more similar to age of reproductive competence.

2.4 Apical Meristem Organization for Reproductive Initials

The dominant part of the seed plant life cycle (Fig. 2.2) is the large, long-lived adult sporophyte as shown in Photo 2.1. At the end of each branch tip is a shoot apex. Some fraction of these apices will produce reproductive cells or initials. The pine

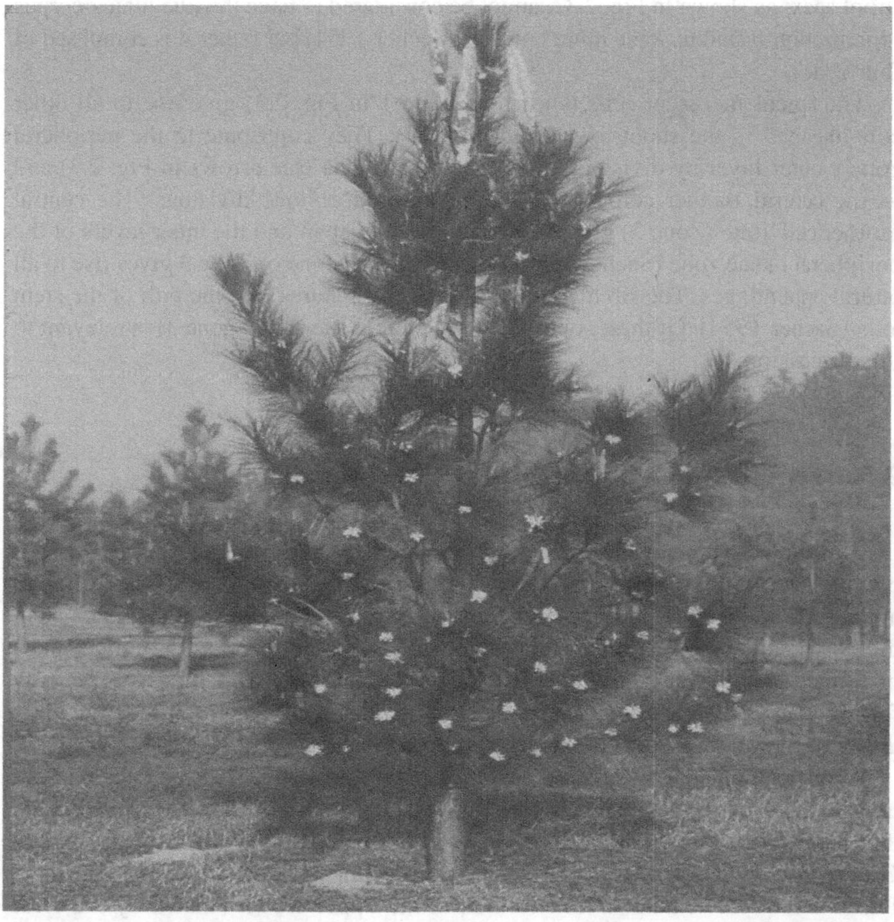

Photo 2.1 Young *Pinus taeda* sporophyte. Male strobili are nearly ready for pollen release and clear bags isolate receptive female strobili located at the top of the crown (Photograph by David Bramlett. Permission granted by his estate)

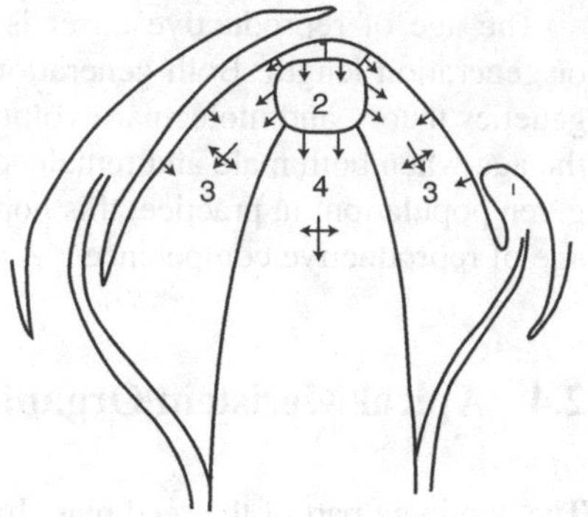

Fig. 2.3 Diagram of *Pinus* spp. during the active growing season. Apical initials are in Zone 1 (*top*). Central mother cells are in Zone 2. Peripheral tissue is Zone 3 and the rib meristem is Zone 4 (From J.A. Sacher (1954) Structure and seasonal activity of the shoot apices of *Pinus lambertiana* and *Pinus ponderosa*. American Journal of Botany 41: 749–759. Copyright permission granted)

shoot apex, as shown in Fig. 2.3, cannot be considered to have the true tunica-corpus organization found in some other conifers (Sacher 1954) but rather it is composed of four zones.

The apical meristem cells (shown as Zone 1 in Fig. 2.3) give rise to all other cell lineages in the shoot apex (Sacher 1954). They contribute to the peripheral zone's outer layer by divisions in the anticlinal plane (see arrows in Fig. 2.3) and to the central mother cell zone by periclinal and oblique divisions. The central mother cell zone (Zone 2) gives rise to the rib meristem and the inner layers of the peripheral tissue zone (Sacher 1954). The peripheral zone or Zone 3 gives rise to all lateral appendages. The rib meristem, or Zone 4, matures into the pith of the stem axis (Sacher 1954). Of these, only the peripheral tissue zone (Zone 3) is relevant to this discussion.

2.5 Female and Male Strobilus Development

In some apices, male and female strobilus initials may form from Zone 3 tissues by late summer. All apices produce shoot growth so each year, it is a different apical cell lineage that gives rise to another set of reproductive initials. Each year, these reproductive cells differentiate into initials which develop into either male or female strobili. Each strobilus, whether male or female, is composed of mostly fertile scales. Next, sporangial tissues form on one side of each fertile scale. Only a few of sporangial cells will undergo sporogenesis (Fig. 2.3).

A consequence of heterospory is the spatial separation between female and male strobili in the same crown (Mitton 1992). This is the case for *Pinus taeda*; the degree of spatial separation increases with age and size of the crown (Fig. 2.5; Greenwood 1980). Little spatial separation occurs in young trees, as shown in Photo 2.1 where female strobili formed in the upper crown (shown by the clear pollination bags) but male strobili formed throughout the crown.

Box 2.1 The adult sporophyte: one genotype or many?

Each apical meristem gives rise to vegetative growth, reproductive initials and next year's apical meristems, so here is the potential for accumulating *de novo* somatic mutations on different branches. Somatic mutations accumulate in different apical meristems or even in different cell lineages within the same apical meristem as shown for some of the Cupressaceae (Korn 2001).

It is thought that some of these mutated cell lineages will give rise to reproductive initials which in turn develop into haploid gametophytes. Steady-state mutational accumulation is assumed to be linear with respect to time (the molecular clock concept) so with advancing age, each cell lineage should accumulate a larger share of somatic mutations. If so, then older perennial plants should have accumulated more mutations in their most recent shoot apices than younger plants. As most spontaneous mutations tend to be mildly deleterious rather than beneficial, this mutational accumulation with advancing age should thus be manifest as a decline in fecundity or perhaps as reduced seed and pollen viability (Drake et al. 1998).

Such a model for aging in perennial plants has far-reaching implications. With time, each perennial plant becomes a living record of its past somatic mutation events. Each apical meristem now has its own historical cadre of mutations and thus each branch can be regarded a separate genotype (Klekowski 1988; Klekowski and Godfrey 1989). This challenges the idea that the aging adult sporophyte is a single totipotent genotype; it is better described as a metapopulation of different genotypes. But do these somatic mutations in the apical meristems actually get passed along through meiosis to gametophytes and gametes?

So far, the answer appears to be no if the process works the same for female and male gametophytes. This early answer comes from a recent experiment with *Pinus strobus* (Cloutier et al. 2003). Their experiment was designed as follows: DNA was extracted from branch tips on two identical clones (or ramets) on a total of 12 different trees. In addition, mature cones collected from three branches on each of the 12 trees provided seeds and from these seeds, female gametophytes were sampled for DNA. Using these DNA samples, mutation rates could be determined using microsatellites, a hypervariable form of DNA polymorphism. Not a single mutation could detected among branches, meristems or female gametophyte tissues in any sample (Cloutier et al. 2003). Each of the 12 adult sporophytes was composed of a single genotype. This surprising result implies the presence of one or more barriers to mutations: stringent DNA repair systems, strong selection at the haploid stages or other barriers yet to be identified. Finding taxon-specific barriers to mutational transmission between sporophyte and gametophyte phases could be a richly rewarding research area.

This can be illustrated by following the pattern of *Pinus taeda* shoot morphogenesis (Greenwood 1980; Figs. 2.4 and 2.5). A single growth cycle (or flush) consists of a stem segment. Along this stem segment is a spiraling series of lateral long shoots. The lateral shoots differentiate into either branches (vegetative) or strobilus initials (reproductive).

In the Northern Hemisphere, female *Pinus taeda* strobili initials first appear in the upper quarter of the crown by late August. By the following spring, they emerge as single female strobili. The male strobilus initials form in clusters on short shoots. Here too, clusters of *Pinus taeda* buds can be seen on the lower quarter of the crown by mid-September even though pollen shed will not take place until the following March (Greenwood 1980).

In older crowns, female strobili form in long-shoot terminal buds (LSTB) in the upper quarter of the crown and male strobili form on the same bud types albeit on branches in the lower quarter of the crown. The LSTB refers to a resting bud composed of an apical meristem flanked by lateral shoot primordia. Female strobili develop on the tips of whorled lateral long shoots (Fig. 2.4; Greenwood 1980) and male strobili in short clusters (Fig. 2.5; Greenwood 1980).

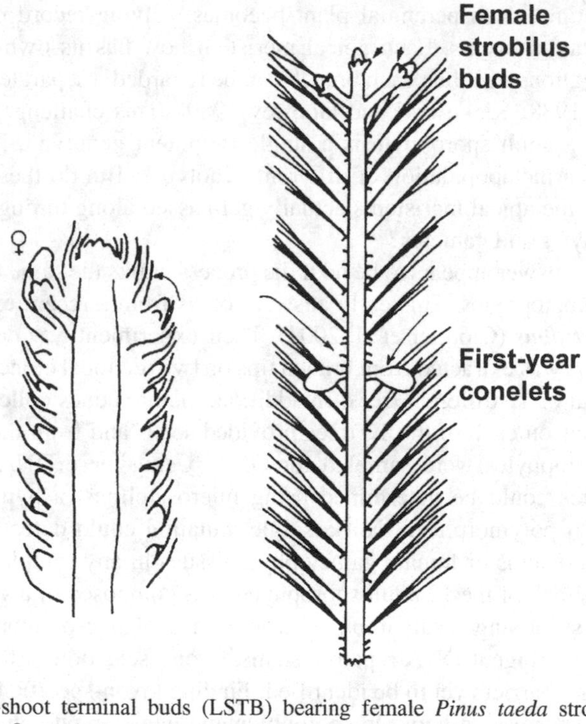

Fig. 2.4 Long-shoot terminal buds (LSTB) bearing female *Pinus taeda* strobilus initials in September (see ♀ symbol) and then the shoot arising from the same bud 17 months later, in February (from Greenwood 1980). Female strobili develop on the tips of whorled lateral branches (From Greenwood M. 1980). Reproductive development in loblolly pine. I. The early development of male and female strobili in relation to long shoot behavior. American Journal of Botany 67: 1414–1422. Copyright permission granted)

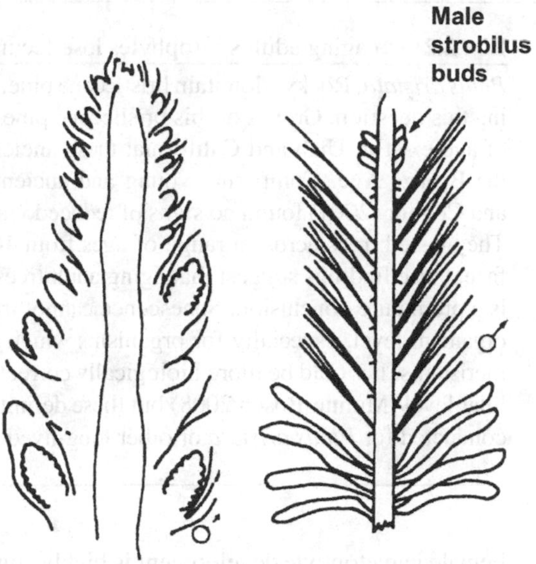

Fig. 2.5 Long-shoot terminal buds (LSTB) bearing male *Pinus taeda* strobilus initials in September and then the shoot arising from the same bud (see ♂ symbol) 17 months later, in February (from Greenwood 1980). Male strobili develop in short clusters. (From Greenwood, M. 1980. Reproductive development in loblolly pine. I. The early development of male and female strobili in relation to the long shoot growth behavior. American Journal of Botany 67:1414–1422. Copyright permission granted)

Reproductive initials, either male or female, form annually in late summer along both peripheral edges (Zone 3 in Fig. 2.3) within LSTB apical meristems (Sacher 1954; Greenwood 1980; Harrison and Slee 1992). Each reproductive initial develops into a strobilus which in turn houses many specialized tissues including sporogenous cells (Fig. 2.2).

2.6 Female Meiosis

These specialized sporogenous cells develop further into megaspore mother cells (MMC) in the female lineage. These undergo meiosis, reducing chromosomal complement by half. Female meiosis is described in more detail in the following chapter.

2.7 Monospory: From Female Megaspore to Gametophyte

The tiny female gametophyte, the size and color of a rice grain, is hidden inside the ovule (endoscopic). It relies on the adult sporophyte for regulatory cues, protection and nutrition (Mable and Otto 1998). Deep inside the megasporangium or ovule, the megaspore mother cell (MMC) gives rise to four megaspores but only one of these develops in a multicellular female gametophyte (monospory).

Box 2.2 Do aging adult sporophytes lose fecundity?

Pinus aristata, Rocky Mountain bristlecone pine, is the ideal species for addressing this question. Groves of this bristlecone pine species exceeding 4,000 years of age exist in Utah and California; these ancient trees were seedlings during the Bronze Age. Comparing young and ancient *Pinus aristata* trees, Lanner and Connor (2001) found no signs of reduced viability in either seeds or pollen. They tested trees across a range of ages from 100 to 4,000 years. On the surface, these findings suggest that aging adult trees do not lose fecundity but this is a premature conclusion. Senescence cannot truly be measured at the whole-organism level, especially for organisms which grow via indeterminate apical meristems. It would be more biologically correct to test for senescence at a cellular level (Munne-Bosch 2008) but these definitive experiments have yet to be conducted for *Pinus aristata* or other long-lived perennial plants.

Female gametophyte development is highly conserved among all gymnosperms. As gymnosperms, conifers have a haploid female gametophyte but in no sense of the term can this be a synonym for an endosperm. Both gymnosperms and angiosperms have a female gametophyte but only angiosperms have an endosperm, the product of multiple fertilization.

Multiple egg cells develop within the female gametophyte. Each is housed in an archegonium. This too is a conserved character; nearly all gymnosperm taxa, including dioecious taxa, have multiple archegonia per ovule.

2.8 Male Meiosis, Microspores and Male Gametophytes

Meiosis occurs in pollen sacs on the underside (abaxial side) of each microsporophyll (Ferguson 1904; Korol et al. 1994). The product of male meiosis is a tetrad of four microspores. The tetrad breaks apart and each of its four microspores develops a pollen wall from tapetal secretions. Together with the pollen wall, the microspore composes the pollen grain. The male gametophyte, now enclosed within a protective pollen wall, undergoes a series of asymmetric mitoses to become multicellular.

Pollen grains are released as the male strobilus desiccates. Released pollen grains are windborne and thus trajectory is subject to the vagaries of air currents. If the pollen lands on a female strobilus and the female strobilus is receptive, then the pollen grain may enter the micropylar opening of an ovule.

The pollen grain, once hydrated, germinates. Its germination tube, with its singular purpose of delivering sperm cells to the egg, penetrates the nucellar tissue of the ovule. This siphonogamous character of the pollen grain is another conserved feature of the life cycle. Conifers also have a long delay between pollen germination and fertilization; this delay can last months or even years. The delay is marked by the development of the female gametophyte and its egg cells.

Box 2.3 Skipping the haploid phase

The Algerian desert conifer *Cupressus dupreziana* omits the haploid phase altogether. Its embryos form directly from diploid, unreduced pollen grains (Pichot et al. 2001). This type of paternal apomixis is so unusual among eukaryotes that this conifer's mating system received mention in the *New York Times*. Science journalist Dr. Olivia Judson reported on this Algerian desert conifer's novel reproduction in her *New York Times* story "Evolving the Single Daddy" printed on September 24, 2008.

Let's start at the beginning. The paternal parent *Cupressus dupreziana* produces pollen grains which are not reduced during meiosis and thus remain diploid. Its pollen grain are captured by a sympatric relative, *Cupressus sempervirens*. The captured pollen grain develops into an embryo inside the ovule of its surrogate mother. Note that these embryos are not hybrids from *Cupressus. sempervirens x Cupressus dupreziana* but rather they have the exact same two haplotypes as their paternal parent (Pichot et al. 2001). To be exact, this mating system is classified as gametophytic apomixis where pollen development circumvents reduction of chromosome number during meiosis I (Richards 1997).

So what molecular marker data might one need to prove this unusual single-daddy mating system really exists for *Cupressus dupreziana*? One needs a background in meiosis, haplotypes and Mendelian inheritance of nuclear DNA markers. Here is a hypothetical example: assume that the surrogate maternal *C. sempervirens* parent is homozygous. It has two chromosomal segments (or haplotypes) which are both A_1-B_6 which means that marker locus A has allele 1 and it is linked to marker locus B which has allele 6.

Next, assume that the paternal *Cupressus dupreziana* parent is heterozygous and that it developed from normal fertilization of two gametes. Its DNA assays might show that this tree inherited a chromosomal segment A_5-B_3 from its father and a chromosomal segment A_7-B_9 from its mother.

None of these four alleles are common to the surrogate *C. sempervirens* tree so now one can use this system to test the single-daddy hypothesis. Let's start with formulating the null hypothesis. If normal male and female meioses were followed by fertilization between haploid male and female gametes then every embryo would have same maternal haplotype of A_1-B_6 plus it would have inherited one of four paternal haplotypes A_5-B_3 or A_7-B_9 (parental or nonrecombinant) or either of the two recombinant haplotypes. But this was not the case. All seeds collected from the maternal parent will have the exact same genotype as the paternal parent: A_5-B_3 /A_7-B_9. No recombinant haplotypes are found among any of these seeds; only the parent's own two haplotypes are here so this provides the proof that meiosis did not reduce the chromosomal complement in the paternal *C. dupreziana* parent. Paternal gametophytic apomixis can be accepted because none of the offspring have the maternal haplotype and all have the same two paternal (nonrecombinant) haplotypes.

(continued)

Box 2.3 (continued)

Perhaps the single-daddy mating system circumvents extinction. After all, *Cupressus dupreziana* is an endangered species on the 2008 IUCN Red List of Threatened Species (www.iucnredlist.org, accessed January 19, 2008). Only a few hundred trees grow along a 12 × 6 km strip along the southwest border of the Tassili Plateau in Algeria and in the Atlas Mountains of Morocco.

During this delay, the pollen tube halts midway through the nucellar tissue. It will resume its growth a few days before fertilization; this pollen tube will eventually deliver its sperm cells or sperm nuclei to the egg cells or archegonia, completing fertilization.

2.9 Syngamy, Fertilization and Seed Maturation

The zygote is the start of a new sporophyte. The zygote matures into an embryo inside the female gametophyte which in turn is housed by the developing seed. Note that a developing seed has all three phases of the diplohaplontic life cycle at one point in time: tissues from the adult sporophyte, the female gametophyte and its young sporophyte (embryo) within. Dispersal of mature embryos is the final defining feature of the seed plant life cycle.

2.10 Closing

The diplohaplontic life cycle of a seed plant has five defining properties: (1) heterospory or separate cell lineages for male and female reproduction, (2) multicellular gametophytes, (3) retention of the female gametophyte within the adult sporophyte, (4) the pollen tube and (5) dispersal of mature embryos. To this, one can add monosporangiate strobili, concurrent diploid and haploid phases, geitonomogamous selfing and iteroparity. In the remaining chapters, I show how these defining characters shape the basic plan or the *Bauplan* of the dynamic wind-pollinated mating system for conifers.

The diplohaplontic life cycle is the frame for the book. The dominant part of the life cycle for conifers is the large, long-lived adult sporophyte and as such, it is concurrent with the annual gametophyte phase. A few of its vegetative meristems on its branches will develop reproductive cells along the peripheral edge of each meristem each year. These cells develop into either female or male strobili which bear sporangial cells. These specialized cells undergo meiosis, reducing chromosomal complement by half. The haploid cells divide and grow into gametophytes, the haploid phase in the life cycle. The peculiar reproductive biology of conifers follows over the next chapters.

References

Bonnet-Masimbert, M. and J. Webber. 1995. From flower induction to seed production in forest tree orchards. Tree Physiology 15: 419–426.

Bramlett, D., C. Williams, et al. 1995. Surrogate pollen induction shortens the breeding cycle of loblolly pine. Tree Physiology 15: 531–575.

Burris, L., C. Williams, et al. 1992. Scion age and its effect on flowering and early selection age. Knoxville, TN, Southern Forest Tree Conference.

Cloutier, D., D. Rioux, et al. 2003. Somatic stability of microsatellite loci in eastern white pine, *Pinus strobus* L. Heredity 90: 247–252.

Coelho, S., A. Peters, et al. 2007. Complex life cycles of multicellular eukaryotes: new approaches based on the use of model organisms. Gene 406: 152–170.

Connor, K. and R. Lanner. 1991. Effects of tree age on pollen, seed and seedling characteristics in Great Basin bristlecone pine. Botanical Gazette 152: 114–122.

Dorman, K. 1976. *The Genetics and Breeding of Southern Pines*. Washington, DC, US Department of Agriculture, Forest Service U.S. Government Printing Office.

Drake J., B. Charlesworth, et al. 1998. Rates of spontaneous mutation. Genetics 148: 1667–1686.

Ferguson, M. 1904. Contributions to the life history of *Pinus* with special reference to sporogenesis, the development of gametophytes and fertilization. Proceedings of the Washington Academy of Science 6: 1–202.

Greenwood, M. 1978. Flowering induced on young loblolly pine grafts by out-of-phase dormancy. Science 201: 443–444.

Greenwood, M. 1980. Reproductive development in loblolly pine. I. The early development of male and female strobili in relation to the long shoot growth behavior. American Journal of Botany 67: 1414–1422.

Greenwood, M. 1984. Phase change in loblolly pine: shoot development as a function of age. Physiologia Plantarum 61: 518–522.

Harrison, D. and M. Slee. 1992. Long shoot terminal bud development and the differentiation of pollen- and seed-cone buds in *Pinus caribaea* var. *hondurensis*. Canadian Journal of Forest Research 22: 1565–1668.

Hofmeister, W. 1851. Vergleichende Untersuchungen der Keimung, Entfaltung und Fruchbiung hoherer Kryptogamen (Moose, Farrn, Equisetceen. Rhizocarpeen und Lycopodiaceen) und der Samenbildung der Coniferen. Leipzig.

Klekowski, E. 1988. Genetic load and its causes in long-lived plants. Trees 55: 195–203.

Klekowski, E. and P. Godfrey. 1989. Ageing and mutation in plants. Nature 340: 389–391.

Korn, R. 2001. Analysis of shoot apical organization in six species of the Cupressaceae based on chimeric behavior. American Journal of Botany 88: 1945–1952.

Korol, A., I. Preygel, et al. 1994. *Recombination Variability and Evolution*. Chapman & Hall, London.

Kozlowski, T. 1971. Growth and development of trees. Vol. 1. *Seed Germination, Ontogeny and Shoot Growth*. Academic, New York.

Lanner, R. and K. Connor. 2001. "Does bristlecone pine senesce?" Experimental Gerontology 36: 675–685.

Mable, B. and S. Otto. 1998. The evolution of life cycles with haploid and diploid phases. BioEssays 20: 453–462.

Matziris, D. 2002. Short note: Hemaphroditism in black pine. Silvae Genetica 51: 130–131.

Mitton, J. 1992. The dynamic mating system of conifers. New Forests 6: 197–216.

Munne-Bosch, S. 2008. Do perennials really senesce? Trends in Plant Science 13: 216–220.

Pharis, R., J. Webber, et al. 1987. The promotion of flowering in forest trees by gibberellin A_4/A_7 and cultural treatments: a review of possible mechanisms. Forest Ecology and Management 19: 65–84.

Pichot, C., M. El-Maataoi, et al. 2001. Surrogate mother for endangered *Cupressus*. Nature 412: 39.

Richards, A. 1997. *Plant Breeding Systems*. Chapman & Hall, London.

Sacher, J. 1954. Structure and seasonal activity of the shoot apices of *Pinus lambertiana* and *Pinus ponderosa*. American Journal of Botany 41: 749–759.
Sax, K. 1962. Aspects of aging in plants. Annual Review of Plant Physiology 13: 489–506.
Singh, H. 1978. *Embryology of Gymnosperms*. Gebruder Borntraeger, Berlin.
Skinner, D. 1992. Ovule and embryo development, seed production and germination in orchard grown control pollinated loblolly pine (*Pinus taeda* L.) from coastal South Carolina. Master's thesis, Department of Biology, University of Victoria, Victoria, BC, 88 pp.
Strasburger, E. 1894. The periodic reduction of the number of chromosomes in the life history of living organisms. Annals of Botany 8: 281–316.
Wareing, P. 1959. Problems of juvenility and flowering in trees. Journal of the Linnaean Society London Botany 56: 282–289.

Section II
Consequences of Heterospory

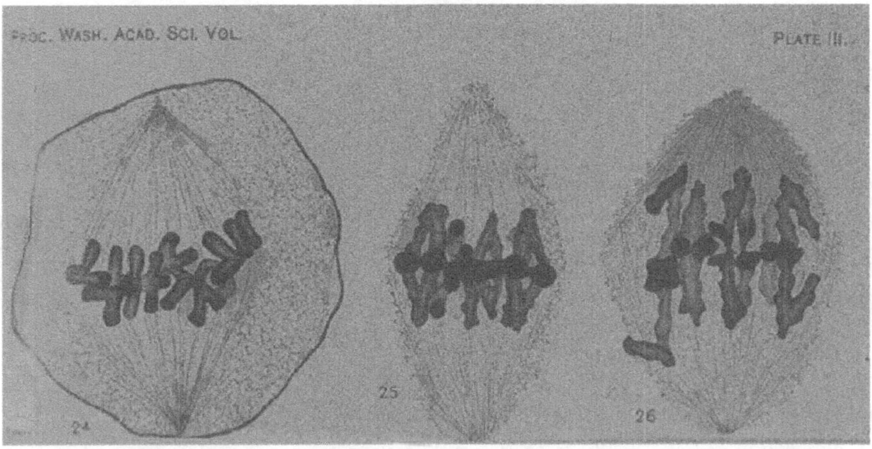

Drawing from M. Ferguson (1904) Contributions to the life history of *Pinus* with special reference to sporogenesis, the development of gametophytes and fertilization. Proceedings of the Washington Academy of Sciences 6: 1–202. Copyright permission granted.

Plate II Male meiosis. This drawing shows the microspore mother cells as follows: drawing (24) *Pinus rigida* equatorial plate stage shows the spindle is pulling towards two opposing poles and drawings (25, 26) metaphase I show outlines of irregular chromosomes for *Pinus strobus*.

Published from Margaret Clay Ferguson's doctoral dissertation at Cornell University, this monograph is the most detailed study of pine meiosis yet to be reported. It contains over 200 hand-drawn figures characterizing fine detail of pine reproduction: meiosis, pollen formation, pollination, pollen tube elongation, female gametophyte development, fertilization and proembryo development. Margaret Ferguson was a professor and then chair of Wellesley's Department of Botany where she conducted her original research on the seed plant life cycle, publishing in *Science* and other scholarly journals.

Chapter 3
Separate Female and Male Meioses

Summary The purpose of this advanced-topics chapter is to examine what is known about meiosis for *Pinus*, other genera within the Pinaceae family and to a lesser extent, other conifers and gymnosperms. Meiosis is the process by which the chromosomal complement is reduced by half. The resulting haploid spores, male and female, develop into multicellular gametophytes. Male and female meioses are both followed by cell division and subsequent growth of multicellular gametophytes. As a general rule, meiosis reshuffles existing genetic variation, generates *de novo* variation and ensures genome stability. But each taxon has its own recombination modification system and here the particulars are presented for *Pinus*, other members of the Pinaceae family and a few close relatives. Strict diploidy is a safe assumption for the Pinaceae. Oddly, most members of the Pinaceae have a uniform karyotype: 12 pairs of metacentric chromosomes. Meiosis I is characterized by few chiasmata per chromosome. Male and female meiosis in *Pinus* spp. are divergent, occurring at different times and producing different tetrad types. Meiotic recombination rates are higher in male gametes than female gametes. An consequence of heterospory, or separate male and female spores, is meiotic divergence between the sexes.

Meiosis has also been so profoundly affected by heterospory that it deserves its own chapter. This transition from homospory, or a single bisexual spore, to heterospory or separate male and female spores, has brought about divergent male and female sporogenesis (Pettit 1970). This can be seen as a change in spore size, increased specialization and even acquired functions such as the siphonogamous pollen tube.

The primary consequence of heterospory is new selection space for divergent male and female meioses as well as respective divergence in gametophyte development. As an example, consider how meiosis has diverged between female and male cells in form, function and timing. Female meiosis II ends in a linear tetrad and from this linear tetrad only one of the four megaspores will survive. But male meiosis II ends with rounded tetrads and from this tetrad, all four microspores will survive. Likewise, one can trace a given fertilization event in conifers back to two widely-spaced meiotic events: the male gamete's meiosis by necessity must occur many months before the female gamete's own meiosis. Disjunct timing between male and female meiosis is another consequence of heterospory; whether it contributes to sex-specific recombination rates remains to be seen.

Meiosis is more than a bridge between diploid and haploid phases of the diplohaplontic life cycle. Its primary function is to reshuffle chromosomal segments into new combinations but it also creates new variation via mutation. Reshuffling occurs by reciprocal DNA exchange. Also known as crossing over, this is the direct result of chiasma formation between homologous chromosomes. New mutations occur by gene conversion or unequal DNA exchange: deletions, duplications and point mutations. Note that this class of meiotic mutations arise do not arise in the same way as somatic mutations.

3.1 Taxon-Specific Recombination Modification Systems

Male meiosis has higher recombination rates than female meiosis. The mechanisms behind sex-specific recombination are not yet known but this is only small part of a larger aggregate of genetic mechanisms shaping meiotic outcome in either sex. This is known as the recombination modification system (Korol et al. 1994). Depending on the taxon, such a system might include transpositional hotspots, heterochromatin knobs, paracentric inversions, supernumerary chromosomes and a host of other genetic mechanisms (Korol et al. 1994; Cai and Xu 2007). For example, recombination rates may differ between male and female meioses in *Zea mays*, *Arabidopsis thaliana* and *Pinus* spp. but the causal mechanisms may not be the same. The recombination modification system is not just sex-specific but also taxon-specific.

The recombination modification system has not been described for any conifer to date (Shepherd and Williams 2008). But one of the first questions for such an inquiry is to ask if conifers are truly diploid. Many seed plants are polyploids yet they are still considered to have a diplohaplontic life cycle. In this next section, we find that for most conifer species, the adult sporophyte is truly diploid.

3.2 Strict Diploidy in the Pinaceae Family

A closer look at *Pinus* karyotypes (Fig. 3.1) shows no recent polyploidy (Sax and Sax 1933; Pederick 1967; Komulainen et al. 2003).

Of the 12 chromosomes in Fig. 3.1, eleven are isobrachial and the 12th is slightly heterobrachial (Saylor 1961; Pederick 1967; Pederick 1970; Dial and Stalter 1980; Ohri and Khoshoo 1986; Doudrick et al. 1995). Some ancient polyploidy has been inferred by comparing chromosomes within an ideotype (Doudrick et al. 1995) and by comparing ribosomal DNA sites (Liu et al. 2003) but this assertion is not fully resolved because past events are difficult to infer without using a large number of molecular cytogenetics labels.

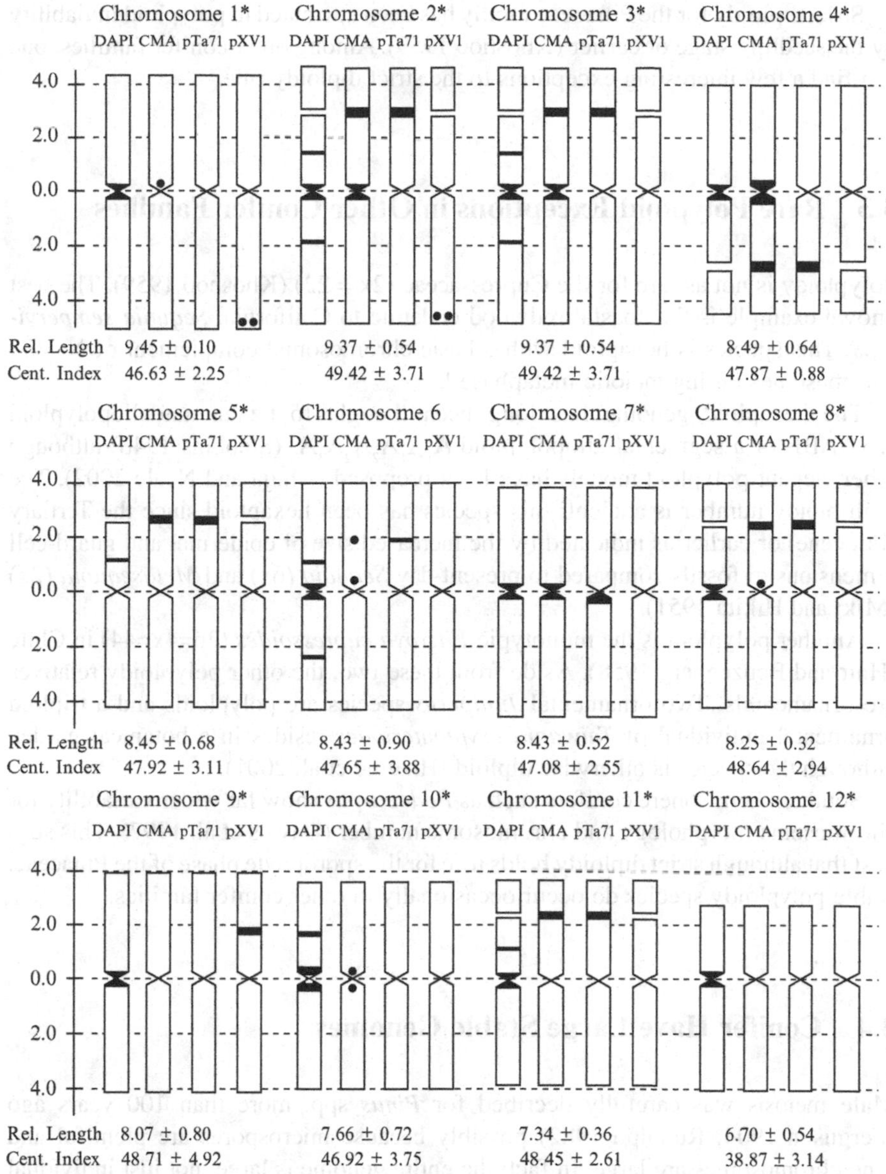

Fig. 3.1 An ideogram of the *Pinus elliottii* karyotype. Darkened sites and bands show the most intense dye signal. Relative lengths of each chromosome are percentages of the total haploid complement. The landmarks are DAPI and CMA, two fluorescent banding dyes, pTa71 is a DNA fragment for 18S-5.8S-25S ribosomal DNA and pXVI is a DNA fragment for the 5S ribosomal DNA (From R. Doudrick et al. (1995) Karyotype of slash pine (*Pinus elliottii* var. *elliottii*) using patterns of influorescence in situ hybridization and fluorochrome banding. Journal of Heredity 86: 289–296) Copyright permission granted

Strict diploidy for the Pinaceae family has been attributed to polyploid inviability by the seedling stage or earlier (Khoshoo 1959). Among other conifer families, one can find a few interesting exceptions to the strict diploidy rule.

3.3 Rare Polyploid Exceptions in Other Conifer Families

Polyploidy is not as rare for the Cupressaceae (2x = 22) (Khoshoo 1959). The best known example is the coastal redwood endemic to California, *Sequoia sempervirens*. This species is hexaploid with a basic chromosomal complement of N = 33 chromosomes during meiotic metaphase I.

The hexaploid genome has long been thought to be an autoallopolyploid AAAABB or a segmental allopolyploid $A_1A_1A_2A_2A_3A_3$ (Stebbins 1948) although other elegant polyploid models have been proposed (Ahuja and Neale 2002). The high ploidy number is ancient; this species has been hexaploid since the Tertiary (Pliocene) or earlier as indicated by the increased size of epidermal and guard-cell dimensions in fossils compared to present-day *Sequoia* (6x) and *Metasequoia* (2x) (Miki and Hikita 1951).

Another polyploid is the monotypic *Fitzroya cupressoides* (2n=4x=44) in Chile (Hair and Beuzenberg 1958). Aside from these two, the other polyploidy relatives are ornamentals. Two ornamental *Juniperus* species are polyploids and a triploid ornamental individual of *Taiwania cryptomeriodes* resides in a botanical garden although the species is otherwise diploid (Hizume et al. 2001).

Southern hemisphere conifers such as *Podocarpus* show far wider variability for chromosome morphology and chromosome number (Davies et al. 1997). This suggest that although strict diploidy holds true for the sporophyte phase of the Pinaceae, stable polyploidy species do occur occasionally in other conifer families.

3.4 Conifer Have Large Stable Genomes

Male meiosis was carefully decribed for *Pinus* spp. more than 100 years ago (Ferguson 1904; Rudolph 1982) possibly because microspores are plentiful and pine chromosomes are large. In fact, the entire genome is large, not just individual chromosomes. Nearly all members of the Pinaceae have uniformly large genome sizes (or C-values) from ranging ~14–32 pg/C (Murray 1998; Wakamiya et al. 1993; Hall et al. 2000; Joyner et al. 2001; Grotkopp et al. 2004) or roughly seven times larger than the human genome.

Meiosis, whether male or female, confers an unusual degree of stability in genome size among the members of the Pinaceae. Genome size is stable even for several F_1 *Pinus* spp. hybrids and their offspring whether crosses are made between closely related allopatric parental species such as *P. elliottii* × *P. caribaea* or distant

Table 3.1 Some of the smaller gymnosperm genome sizes. Note that *Gingko biloba* is among the smaller gymnosperm genomes

Species	Haploid DNA per nucleus (pg/C)	References
Gingko biloba	9.95	Leitch et al. 2001
Araucaria cookie	9.55	Ohri and Khoshoo 1986
Wollemia nobilis	13.94	Hanson 2001
Callitris glauca	8.25	Ohri and Khoshoo 1986
Agathis australis	15.80	Davies et al. 1997
Juniperus virginiana	10.82	Hizume et al. 2001
Thuja plicata	12.84	Hizume et al. 2001
Taxodium mucronatum	8.75	Ohri and Khoshoo 1986
Sequoiadendron giganteum	9.93	Hizume et al. 2001
Metasequoia glyptostroboides	11.04	Hizume et al. 2001
Sciadopitys verticillata	20.80	Hizume et al. 2001

crosses are made between North American and Asian soft pines within the same subsection (Sax 1960; Williams et al. 2002).

Other conifers and gymnosperms have smaller genome sizes than the Pinaceae as a rule (Table 3.1). Monotypic *Gingko biloba* has one of the smallest genomes among the gymnosperms but it also has atypical chromosomal landmarks corresponding to sex determination in this dioecious species (Leitch et al. 2001; Lee 1954) which may constrain increases in genome size (Box 3.1).

Box 3.1 Chromosomal landmarks for sex determination

Gingko biloba is a dioecious gymnosperm with twelve pairs of chromosomes. The sex of the tree can be determined from its karyotype using the presence or absence of a satellite DNA landmark on this small subterminally constricted chromosome. Male trees have only one chromosome with the satellite DNA landmark; if both chromosomes in the pair have the satellite DNA then the tree is female (Lee 1954). This information is valued for culling female trees for the horticultural market. Male and female *Gingko biloba* trees appear identical until reproduction when males produce only pollen and female trees produce fleshy, stinking seeds rich in rancid butyric acid.

3.5 Supernumerary Chromosomes

Some *Picea* spp. are reported to have supernumerary B chromosomes; the number of B chromosomes per cell varies from zero to five among *Picea glauca* and *Picea sitchensis* karyotypes (Moir and Fox 1977). These extrachromosomal elements cause aberrant meiosis in other seed plants but this effect has not been tested for these conifer species. Highly heterchromatic, B chromosomes do not code for genes. They are smaller than autosomal or A chromosomes and do not recombine with the larger complement of paired chromosomes (Jones and Houben 2003); inheritance is autonomous. Whether supernumerary chromosomes should be considered to be part of the recombination modification system for *Picea* spp. or other conifers has yet to be determined.

3.6 Stages of Meiosis

What little else is known about conifer meiosis is given below; the study of plant meiosis has been largely restricted to model organisms such as yeast, maize and *Arabidopsis*. Reciprocal exchange, also known as crossing over, is the most prevalent mechanism in eukaryotes. During meiosis I this begins with chiasma formation and then is followed by equal DNA exchange between homologous chromosomes (Sybenga 1975). Meiosis II closely parallels mitosis.

For the benefit of later discussion, meiosis basics are given here. Meiosis I is divided into prophase I, metaphase I, anaphase I and telophase I (Table 3.2). Of these, prophase I is the most complex and central to this discussion. It has additional sub-stages: leptotene, zygotene, pachytene, diplotene and diakinesis

Table 3.2 Meiosis I stages. This is the reductional division of meiosis. Meiosis II is essentially mitosis as haploid cells duplicate in this part of meiosis. Meiosis I and II produce four haploid cells or spores. From micro- and megaspores develop gametophytes. Stages 1–7 refer to the classification system in Singh (1978, pp. 24–31)

Stages		Description
Prophase I	Leptotene	Chromosomes are long thin threads
	Zygotene (Stages 1–2)	Chromosomes appear as two complete sets. Time of active pairing or synapsis of homologs. Formation of synaptonemal complex, a protein structure, between homologs
	Pachytene (Stage 3)	Full synapsis Nucleoli are prominent
	Diplotene-Diakinesis (Stages 4–5)	Synapsed structure now appears as a bundle of four homologous chromatids. Pairs begin to separate and cross-shaped structures or chiasmata appear. At least one chiasma is essential for proper segregation. A chiasma is a manifestation of reciprocal exchange
Metaphase I	Stage 6	Paired homologs appear on equatorial plane. Centromeres do not divide. The two centromeres attach to spindle fibers from opposite poles
Anaphase I		Chromosomes move directionally to the poles. Members of a homologous pair each move to an opposite pole
Telophase I	Stage 7	Two haploid nuclei appear

(Table 3.2). Pairing configurations and the number of chiasmata can be seen at certain stages of meiosis I (Ferguson 1904). Chiasma are conserved in position from the time of their first appearance at early diplotene until late metaphase I (Sax 1932). Also included here are stages standardized for pollen development (Singh 1978, pp. 24–31).

3.7 Chromosomal Segregation as a Source of Variation

Chromosomal segregation alone contributes to new combinations of alleles if numbers of haploid chromosomes are high. Segregation refers to independent sorting of the four chromatids into different spore cells upon completion of meiosis. Each spore cell, either microspore or megaspore, receives a haploid chromosomal complement. As a rule, the number of different chromatid combinations is 2^n where n is the haploid chromosome number.

For example, the haploid chromosome number for *Pinus* spp. is 12 so the number of possible combinations from segregation now increases to 2^{12} or 4,096 different chromatid combinations. Aside from the additional variation from DNA exchange, the process of chromosome segregation itself (without crossovers) actually shuffles new combinations among microspore and megaspore cells. The number of new combinations is even greater when any two gametes form a zygote.

3.8 Reciprocal DNA Exchange During Prophase I

Meiosis in the Pinaceae proceeds with an unusually high degree of uniformity yet few chiasmata are evident for these large chromosomes. Bivalents form regularly; univalents or multivalents are rare. Chiasma frequency for different genera within the Pinaceae is surprisingly similar, varying between 1.9 and 2.7 chiasmata per bivalent. For example, *Picea abies* has 2.7 chiasmata, *Larix* has 2.4 chiasmata and various *Pinus* species have 2.3–2.5 chiasmata (Sax 1932).

The range of chiasmata per cell is quite narrow. Only one to four chiasmata are observed per cell for *Pinus* spp. and *Larix* spp. (Ferguson 1904; Sax 1932, 1960; Sax and Sax 1933; Kedharnath and Upadhaya 1967). As a specific example, *Larix decidua* has chiasma frequency ranging from one to four per bivalent with an average of 2.36. Only four percent of the cells have one chiasma, 63% have two chiasmata, 25% have three and only 8% have four (Sax 1932). These data suggest reciprocal DNA exchange is low for these large chromosomes.

Some evidence suggests that chiasma position might be localized and conserved in position from the time of first appearance at early diplotene until late metaphase I (Sax 1932). If so, this might be one of the means by which karyotypic orthoselection (Box 3.2) could be operative in conifer genomes.

Box 3.2 Is karyotypic orthoselection operative during meiosis?

Strict diploidy, karyotypic similarity and similar genome sizes together characterize members of the Pinaceae family. These features have been long viewed as evidence for a slow rate of evolutionary divergence (i.e. Prager et al. 1976; Levin and Wilson 1976). But karyotypic orthoselection is another explanation yet to be ruled out. This refers to the active process of selecting against genomic divergence. Here, a strong selective force conserves the genome, blocking any change to basic chromosome number or karyotypic similarity (Brandham and Doherty 1998). Karyotypic orthoselection could be stringent enough to restrict large-scale chromosomal rearrangements for the Pinaceae even though small-scale chromosomal rearrangements such as paracentric inversions do occur (Saylor and Smith 1966; Shepherd and Williams 2008).

A meiotic mechanism selecting for diploidy was hypothesized by Hally Sax in 1932 although it has yet to be tested. Noting that *Larix* spp. has 12 pairs of metacentric chromosomes, she observed meiosis I for two *Larix* species and their fertile adult F1 hybrid. In all three cases, meiotic divisions were uniform to such an extent that unpaired chromosomes (univalents) were rarely observed. Both parental species and F1 hybrid all had an average of 2.4 chiasmata per bivalent, attesting to the unusual genome stability during meiosis. But the most interesting finding was that chiasmata were localized. These occurred mostly within the interstitial regions of each chromosome. This is the region between the median or subterminal part of the chromosome, located between the spindle attachment and the distal end of the chromosome. Localized chiasma formation is an important finding because it is a requisite condition for her proposed meiotic mechanisms selecting for strict diploidy (Sax 1932).

Her reasoning was this: if localized chiasmata could form at the ends of the chromosome, then the movements of a polyploid's quadrivalents would not be constrained and they would be less likely to break into aneuploid fragments (Sax 1932). Polyploid gametes would survive because they would have a full chromosomal complement. But if the localized chiasmata can only form within the interstitial region, then this is not the case and aneuploidy would result, causing inviable gametes (Sax 1932). This intriguing hypothesis deserves a closer look using molecular cytogenetics. Inferences from low-density *Pinus taeda* genetic mapping data do not appear to support this finding (Zhou et al. 2003; Williams and Reyés-Valdés 2007) but more research is required.

Evidence favoring the karyotypic orthoselection hypothesis is empirical. One source of support is the unusual degree of karyotypic similarity among the members of the Pinaceae family. Nearly all 200 members within the Pinaceae family have 12 pairs (2x = 24) of metacentric chromosomes

(continued)

Box 3.2 (continued)

(Chamberlain 1899; Ferguson 1904; Sax and Sax 1933). The only exception is *Pseudotsuga menziesii* which has 13 pairs of chromosomes (2x = 26) even though all other *Pseudotsuga* species have only 12 pairs (2x = 24) (El-Kassaby et al. 1983). This karyotypic similarity is ancient, persisting in both hard and soft pine subgenera which diverged 136–190 million years ago (MacPherson and Filion 1981).

Karyotypic similarity for the Pinaceae is unusually high when compared to many angiosperm families. While a few angiosperm taxa such as *Lathyrus*, *Vicia*, *Allium*, and *Vigna* species do have karyotypic similarity (Raina and Rees 1983; Narayan 1988; Parida et al. 1990), it rarely extends to the family level (Brandham and Doherty 1998). More typical for angiosperms are gross chromosomal rearrangements or changes in base chromosome number. Many angiosperm families typically show five and 15 chromosomal rearrangements on short time scales ranging from 10 and 24 MY (Gaut and Doebley 1997; Lagercrantz 1998; Paterson 1996; Reinisch et al. 1994; Tanksley et al. 1992; Wendel 1989; Whitkus et al. 1992).

3.8.1 Female Meiosis: A Case of Monospory

Rate of crossing over or the recombination fraction, was shown to be higher for male meioses in *Pinus radiata* but this was based only on a single map interval (Moran et al. 1983). But larger mapping efforts in *Pinus taeda* bore out this finding; recombination rates were 26% (Groover et al. 1995) and this was the case for male meioses in *Pinus pinaster* (Plomion and O'Malley 1996). Female recombination rates are lower than those for males.

One minor factor might be a difference in the environmental conditions during meiosis itself. Consider that for a given fertilization event, the respective male and female gametes come from temporally separated meiotic events. The male meiotic event leading to the pollen grain occurred many months prior pollination but the female meiotic event leading to the egg cell occurred closer to pollination.

In *Pinus taeda*, female meiosis occurs in the late winter or early spring just prior to fertilization. Specific dates have been reported for *Pinus taeda* megasporogenesis (Skinner 1992). A few megaspore mother cells first appear in late January in coastal South Carolina USA but meiosis is not synchronized within a female strobilus or within a single tree. Enlarging MMC and onset of prophase I are present by mid-March (Skinner 1992). Linear tetrads of megaspores fully form by March 24 but by April 21, meiosis is now complete and degenerating megaspores are no longer seen (Skinner 1992).

Another factor which might lower recombination rates for female meiosis is that the surviving megaspore depends on tetrad position; the megaspore closest to the chalazal end of the ovule survives (Skinner 1992). Do the chromatids leading to the formation of this megaspore have same pattern of chiasma formation as the other three? This is not yet known; female meiosis has not been well-studied in conifers. This single megaspore undergoes somatic growth to become a large, translucent female gametophyte.

3.8.2 *Male Meiosis Produces Four Microspores*

Meiosis occurs in pollen sacs on the underside of each microsporophyll in late winter, before pollination. For *Pinus taeda*, the timing of male meiosis begins in late January, only 3–4 weeks before the February–March pollen flight in southern Mississippi USA (Mergen et al. 1963). The end product of male meiosis is a tetrad of four microspores. The tetrad breaks apart and each of its four microspores develops a pollen wall from tapetal secretions. The pollen wall and microspore together composes the pollen grain. The male gametophyte, enclosed within a protective coating provided by the adult sporophyte, now undergoes mitoses to becomes multicellular.

3.9 From Sex-Specific Meiotic Events to Genetic Mapping

Using two or more linked DNA markers, one can assay DNA from related individuals (usually sibs) then count the number of recombinant haplotypes recovered relative to those with parental haplotypes. This recombination fraction can be directly counted from DNA assays of haploid gametophyte tissue. A single pollen grain can be assayed with some difficulty (Kostia et al. 1995) but the female gametophyte is a more apparent choice because it is the identical, haploid genetic complement to the egg nucleus and produces abundant tissue for DNA extraction (Carlson et al. 1991).

Another method is to infer past recombination events using DNA from diploid descendants. Given a map, then one can trace a founder's segments across generations (Box 3.3). When male and female meioses have different recombination rates, then one must use a sex-averaged genetic map (Gwaze et al. 2003) as shown in Fig. 3.2.

Recovery of male and female meiotic products for this particular *Pinus taeda* map was indirectly estimated using both haploid and diploid tissues for a full-sib group and codominant marker systems with multiple alleles; more saturated maps now exist for sophisticated mapping methodology.

Two DNA markers are considered linked if they reside on the same chromosome and thus co-segregate during meiosis. Each chromosome segregates independently

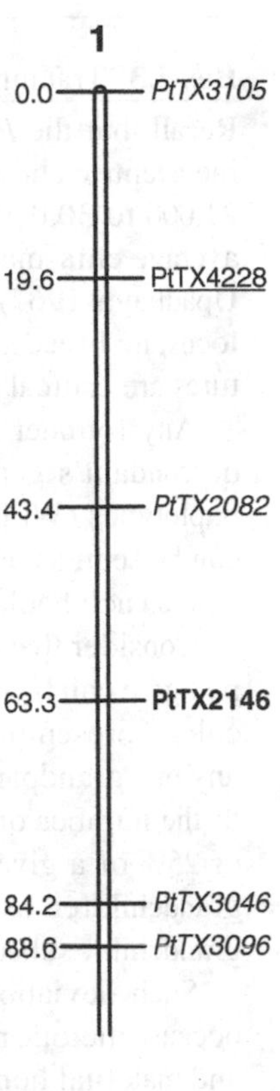

Fig. 3.2 Linkage 1 group from the sex-averaged *Pinus taeda* map corresponds to part of a chromosome. On the *right* side, linked DNA microsatellite markers are shown. On the *left*, recombination fractions have been converted into genetic distances in centimorgans (cM-Kosambi) (From Zhou, Y., D. Gwaze et al. 2003. No clustering for linkage map based on low-copy and undermethylated microsatellites. Genome 46: 809–816. Copyright permission granted)

during meiosis so DNA markers residing on different chromosomes will have alleles which do not co-segregate and these will be considered unlinked. Chromosome segments, entire chromosomes or entire sets of chromosomes can be defined as haplotypes depending on distribution and number of linked DNA markers.

Genetic maps can be constructed from many types of DNA polymorphisms. These markers correspond to one or more sites in the genome but they vary with respect to number of alleles, binding sites within a genome and ease of use. Some individuals in the mapping population will show intervals marked by no crossovers; the crossover events in this case fall outside the regions defined by the DNA markers. In these cases, the original paternal or maternal haplotype is intact and these are defined as parental haplotypes.

If a crossover occurs within the interval between two linked markers then various types of crossover haplotypes are recovered. The more chiasmata per chromosome pair, more classes of crossover haplotypes that can be recovered. This simple example illustrates that measuring meiotic events serves as the foundation for constructing a genetic map with polymorphic markers. Fractions for each class of recombinant and parental haplotypes are tallied then translated into genetic map distances using one of several map functions. The best choice of a map function depends on which set of biological assumptions best fit meiotic recombination itself.

Box 3.3 Tracing a founder's chromosomal segments

Recall that the *Pinus* spp. karyotype is strictly diploid with only 12 pairs of metacentric chromosomes, it has a colossal nuclear genome size ranging from 21,000 to 30,000 Mb per haploid nucleus (Wakamiya et al. 1993) and yet its average chiasma frequency is only 2.3–2.5 (Ferguson 1904; Kedharnath and Upadhaya 1967). This species has high heterozygosity, multiple alleles per locus, no breed structure and little population differentiation. All of these features are critical to tracing each founder's genomic contribution.

Any founder's genomic contribution, if known, can be inferred within a descendant's genome by identifying intact chromosome segments or DNA haplotypes (Williams and Reyés-Valdes 2007). The genome of an individual can be seen as a mosaic composed of chromosomal segments (or haplotype) where each haplotype is inherited from a different founder.

Consider that a grandchild's genome within a three-generation *Pinus taeda* pedigree can be partitioned into a four-color mosaic of DNA haplotypes, each color representing one of its four grandparents. The expected contribution of any one grandparent to its entire cohort of grandchildren is $½^n$ where n refers to the number of generations between the founder and its descendant so $½^2$ or 25% of a given grandparent's genome was transmitted *on average* to the grandchildren as a group assuming Mendelian inheritance. But for any one grandchild, substantial deviations occur from this expected value.

Such deviations are expected within the genome of a single individual because meiotic recombination is an uneven process. During meiosis, paternal and maternal homologs are joined by chiasmata which mark reciprocal DNA exchange between homologs, or crossover events. Each meiotic event has chiasmata appearing at different positions along a given homolog. Numbers and position of chiasmata vary although orderly disjunction requires no less than one chiasma or crossover during meiosis I.

Each gamete thus receives a different set of haplotypic segments from its founders. No two descendant genomes will have the same array of haplotypic segments because each descendant genome is the union of two independent meiotic products or gametes, one from each parent. If each crossover position is independent from the next then interference can be assumed to be negligible and crossover positions along chromosome follow a Poisson process (Haldane 1919). A founder's haplotypic segments, measured in map units, will be recovered at different positions and in different amounts for each descendant's genomes (Mather and Jinks 1971, p. 326).

Measuring such founder proportions was largely theoretical until the advent of high-throughput DNA sequencing data. Estimating founder contributions began with Fisher's (1949, 1953) theory of junctions. The theory of junctions tracks parental chromosomal blocks. Each chromosome is viewed as a genomic continuum composed of junctions and tracts, not as a set of discrete marker points.

(continued)

Box 3.3 (continued)

Each junction is a crossover event identified in relation to its chromosomal landmarks. The position of crossover events thus delineates the tracts or blocks inherited from different grandparents or founders. Building on the theory of junctions, Donnelly (1983) derived a hypercube method for determining the probability that a founder contributed none of its DNA segments to a distant descendant. This was later extended in the concept of *identity-by-descent proportions* or IBDP by Guo (1994, 1995) and others (Bickeböller and Thompson 1996; Stefanov 2000; Baird et al. 2003).

The IBDP value is a random variable moving along the genomic continuum (Guo 1994); it is not a single value such as the identity-by-descent (IBD) probability. Use of IBDP is confined to small sets of relatives ($N < 8$) and thus not amendable to large pedigrees or highly heterozygous populations (Maliepaard et al. 1997; Williams 1998).

This method has been further extended to complex highly heterozygous outbred pedigrees or even breeds (Haley et al. 1994) and a computational shortcut is now available (Reyés-Valdes and Williams 2002). Such a shortcut becomes useful for tracing founder segments for outcrossing species such as *Pinus taeda* because tracing chromosomal segments from each founder (Williams and Reyés-Valdes 2007) becomes quite complex and computationally demanding for conifers as well as any other outcrossing organism whether wild, captive or domesticated.

3.10 Closing

This chapter illustrates how heterospory has led to divergent female and male meiosis. Although meiosis has been well-documented for 100 years in *Pinus* spp., the recombination modification system has not yet been fully characterized. Strict diploidy, high karyotypic similarity and stable large genome sizes suggest karyotypic orthoselection is likely to be operative. Other anomalies which could potentially affect meiotic fidelity but have not yet been well-studied include the following: effect of retrotransposition, prevalence of DNA repair systems, B chromosomes and illegitimate recombination. From this point forward, the book's section is organized around heterospory and the diplohaplontic life cycle of seed plants. Here the reader will find more detailed accounts of the peculiar reproductive biology of the Pinaceae.

References

Ahuja, M. and D. Neale. 2002. Origins of polyploidy in coast redwood (*Sequoia sempervirens* (D.Don.) Endl.) and relationship of coast redwood to other genera of Taxodiaceae. Silvae Genetica 51: 93–100.

Baird, S., N. Barton, et al. 2003. The distribution of surviving blocks of an ancestral genome. Theoretical Population Biology 64: 451–471.

Bickeböller, H. and E. Thompson. 1996. The probability distribution of the amount of an individual's genome surviving to the next generation. Genetics 143: 1043–1049.

Brandham, P. and M.-J. Doherty. 1998. Genome size variation in the Aloaceae, an angiosperm family displaying karyotypic orthoselection. Annals of Botany 82: 67–73.

Cai, X. and S. Xu. 2007. Meiosis-driven genome variation in plants. Current Genomics 8: 151–161.

Carlson, J., L. Tulsieram, et al. 1991. Segregation of random amplified DNA markers in F1 progeny of conifers. Theoretical and Applied Genetics 83: 194–200.

Chamberlain, C. 1899. Oogenesis in *Pinus Laricio*. Botanical Gazette 27: 268–281.

Davies, B., I. O'Brien, et al. 1997. Karyotypes, chromosome bands and genome size variation in New Zealand endemic gymnosperms. Plant Systematics and Evolution 208: 169–185.

Dial, S. and R. Stalter. 1980. The karyotype of *Pinus glabra*. Journal of Heredity 71: 297.

Donnelly, K. 1983. The probability that related individuals share some section of the genome identical by descent. Theoretical Population Biology 23: 34–64.

Doudrick, R., J. Heslop-Harrison, et al. 1995. Karyotype of slash pine (*Pinus elliottii var. elliottii*) using patterns of fluorescence in situ hybridization and flurochrome banding. Journal of Heredity 86: 289–296.

El-Kassaby, Y., A. Colangeli, et al. 1983. Numerical analysis of karyotypes in the *Pseudotsuga* genus. Canadian Journal Botany 61: 536–544.

Ferguson, M. 1904. Contributions to the life history of *Pinus* with special reference to sporogenesis, the development of gametophytes and fertilization. Proc. Wash. Acad. Sci. 6: 1–202.

Fisher, R. 1949. *The Theory of Inbreeding*. Academic, New York.

Fisher, R. 1953. A fuller theory of 'junctions' in inbreeding. Heredity 8: 187–198.

Gaut, B. and J. Doebley. 1997. DNA sequence evidence for the segmental allotetraploid origin of maize. Proceedings of the National Academy of Sciences USA 94: 6809–6814.

Groover, A., C. Williams, et al. 1995. Sex-related differences in meiotic recombination frequency in *Pinus taeda* L. Journal of Heredity 86: 157–158.

Grotkopp, E., M. Rejmánek, et al. 2004. Evolution of genome size in pines (*Pinus* spp.) and its life-history correlates: supertree analyses. Evolution 58: 1705–1729.

Guo, S. 1994. Computation of identity by descent proportions shared by two siblings. American Journal of Human Genetics 54: 1104–1109.

Guo, S. 1995. Proportion of genome shared identical by descent by relatives: concept, computation, and application. American Journal of Human Genetics 56: 1468–1476.

Gwaze, D., Y. Zhou, et al. 2003. Haplotypic QTL mapping in an outbred pedigree. Genetical Research 81: 43–50.

Hair, J. and E. Beuzenberg. 1958. Chromosomal evolution in the Podocarpaceae. Nature 181: 1584–1586.

Haldane, J. 1919. The combination of linkage values and the calculation of distances between the loci of linked factors. Journal of Genetics 8: 299–309.

Haley, C., S. Knott, et al. 1994. Mapping quantitative trait loci in crosses between outbred lines using least squares. Genetics 136: 1195–1207.

Hall, S.E., W. Dvorak, et al. 2000. Flow cytometric analysis of DNA content for tropical and temperate New World pines. Annals of Botany 86: 1081–1086.

Hanson, L. 2001. Chromosome number, karyotype and DNA C-value of the Wollemi pine (*Wollemia nobilis,* Araucariaceae). Botanical Journal of the Linnaean Society 135: 271–274.

Hizume, M., T. Kondo, et al. 2001. Flow cytometric determination of genome size in the Taxodiaceae, Cupressaceae *sensu stricto* and Sciadopityaceae. Cytologia 66: 307–311.

Jones, N. and A. Houben. 2003. B chromosomes in plants: escapees from the A chromosome genome. Trends in Plant Science 8: 417–423.

Joyner, K., X.-R. Wang, et al. 2001. DNA content for Asian pines parallels New World relatives. Can. J. Bot. 79: 192–191.

Kedharnath, S. and L. Upadhaya. 1967. Chiasma frequency in *Pinus roxburghii* Sarg. and *P. elliottii* Englem. Silvae Genetica 16: 112–113.

Khoshoo, T. 1959. Polyploidy in gymnosperms. Evolution 13: 24–29.

Komulainen, P., G. Brown, et al. 2003. Comparing EST-based genetic maps between *Pinus sylvestris* and *Pinus taeda*. Theoretical and Applied Genetics 107: 667–678.

Korol, A., I. Preygel, et al. 1994. *Recombination Variability and Evolution*. Chapman & Hall, London.

Kostia, S., S.-L. Varvio, et al. 1995. Microsatellite sequences in a conifer, *Pinus sylvestris*. Genome 38: 1244–1248.

Lagercrantz, U. 1998. Comparative mapping between *Arabidopsis thaliana* and *Brassica nigra* indicates that Brassica genomes have evolved through extensive genome replication accompanied by chromosome fusions and frequent rearrangements. Genetics 150: 1217–1228.

Lee, C. 1954. Sex chromosomes in *Gingko biloba*. American Journal of Botany 41: 545–549.

Leitch, I., L. Hanson, et al. 2001. Nuclear DNA C-values complete familial representation in gymnosperms. Annals of Botany 88: 843–849.

Levin, D. and A. Wilson 1976. Rates of evolution in seed plants: net increase in diversity of chromosome numbers and species numbers through time. Proceedings of the National Academy of Sciences USA 73: 2086–2090.

Liu, Z.-L., D. Zhang, et al. 2003. Chromosomal localization of 5S and 18S-5.8S-25S ribosomal DNA sites in five Asian pines using fluorescence in situ hybridization. Theoretical and Applied Genetics 106: 198–204.

MacPherson, P. and W. Filion. 1981. Karyotype analyis and the distribution of constitutive heterochromatin in five species of *Pinus*. Journal of Heredity 72: 193–198.

Maliepaard, C., J. Jansen, et al. 1997. Linkage analysis in a full-sib family of an outbreeding plant species: overview and consequences for applications. Genetical Research 70: 237–250.

Mather, K. and J. Jinks. 1971. *Biometrical Genetics*. Cornell University Press, Ithaca, NY.

Mergen, F., G. Stairs, et al. 1963. Microsporogenesis in *Pinus echinata* and *Pinus taeda*. Silvae Genetics 12: 127–129.

Miki, S. and S. Hikita 1951. Probable chromosome number of fossil *Sequoia* and *Metasequoia* found in Japan. Science 113: 3–4.

Moir, R. and D. Fox. 1977. Supernumerary chromosome distribution in provenances of *Picea sitchensis* (Bong.) Carr. Silvae Genetica 26: 26–33.

Moran, G., J. Bell, et al. 1983. Greater meiotic recombination in male vs. female gametes in *Pinus radiata*. Journal of Heredity 74: 62.

Murray, B. 1998. Nuclear DNA amounts in gymnosperms. Annals of Botany 82(Suppl A): 3–15.

Narayan, R. 1988. Constraints upon the organization and evolution of chromosomes in *Allium*. Theor. Appl. Genet. 75: 319–329.

Ohri, D. and T. Khoshoo. 1986. Genome size in gymnosperms. Plant Systematics and Evolution 153: 119–132.

Parida, A., S. Raina, et al. 1990. Quantitative DNA variation between and within chromosome complements of *Vigna* species (Fabaceae). Genetica 82: 125–133.

Paterson, A. 1996. *Genome Mapping in Plants*. Academic, San Diego, CA.

Pederick, L. 1967. The structures and identification of chromosomes of *Pinus radiata* D. Don Silvae Genetica 16: 69–77.

Pederick, L. 1970. Chromosome relationships between *Pinus* species. Silvae Genetica 19: 171–180.

Pettit, J. 1970. Heterospory and the origin of the seed habit. Biological Reviews 45: 401–415.

Plomion, C. and D. O'Malley. 1996. Recombination rate differences for pollen parents and seed parents in *Pinus pinaster*. Heredity 77: 341–350.

Prager, E., D. Fowler, et al. 1976. Rates of evolution in conifers (Pinaceae). Evolution 30: 637–649.

Raina, S. and H. Rees. 1983. DNA variation between and within chromosome complements of *Vicia* species. Heredity 51: 335–346.

Reinisch, A., J. Dong, et al. 1994. A detailed RFLP Map of Cotton, *Gossypium hirsutum* X *G. barbadense* – Chromosome organization and evolution in a disomic polyploid genome. Genetics 138: 829–847.

Reyés-Valdes, M. 2000. A model for marker-based selection in gene introgression breeding programs. Crop Science 40: 91–98.

Reyés-Valdes, M. and C. Williams. 2002. A haplotypic approach to founder-origin probabilities and outbred QTL analysis. Genetical Research 80: 231–236.

Rudolph, E. 1982. Women in the nineteenth century American botany: a generalized unrecognized constituency. American Journal of Botany 69: 1346–1355.

Sax, H. 1932. Chromosome pairing in *Larix* species. Journal of the Arnold Arboretum 13: 368–373.

Sax, K. 1960. Meiosis in interspecific pine hybrids. Forest Science 6: 135–138.

Sax, K. and H. Sax. 1933. Chromosome number and morphology in the conifers. Journal of the Arnold Arboretum 65: 356–374.

Saylor, L. 1961. A karyotypic analysis of selected species of *Pinus*. Silvae Genetica 10: 77–84.

Saylor, L. and B. Smith. 1966. Meiotic irregularity in species and interspecific hybrids of *Pinus*. American Journal of Botany 53: 453–468.

Shepherd, M. and C. Williams. 2008. Comparative mapping among subsection *Australes* (genus *Pinus*, family Pinaceae). Genome 51: 320–331.

Singh, H. 1978. *Embryology of Gymnosperms*. Gebruder Borntraeger, Berlin.

Skinner, D. 1992. Ovule and embryo development, seed production and germination in orchard grown control pollinated loblolly pine (*Pinus taeda* L.) from coastal South Carolina. Master's Thesis, Department of Biology. Victoria, BC, University of Victoria: 88.

Stebbins, G. 1948. The chromosomes and relationships of *Metasequoia* and *Sequoia*. Science 108: 95–98.

Stefanov, V. 2000. Distribution of genome shared identical by descent by two individuals in the grandparent-type relationship. Genetics 156: 1403–1410.

Sybenga, J. 1975. *Meiotic Configurations*. Springer, Berlin.

Tanksley, S., M. Ganal, et al. 1992. High density molecular linkage maps of the tomato and potato genomes. Genetics 132: 1141–1160.

Wakamiya, I., R. Newton, et al. 1993. Genome size and environmental factors in the genus *Pinus*. American Journal of Botany 80: 1235–1241.

Wendel, J. 1989. New World tetraploid cottons contain Old World cytoplasm. Proceedings of the National Academy of Sciences USA 86: 4132–4136.

Whitkus, R., J. Doebley, et al. 1992. Comparative genome mapping of sorghum and maize. Genetics 132: 1119–1130.

Williams, C. 1998. QTL mapping in outbred plants. pp. 81–94, Chapter 5. Editor: A. Paterson. In: *Molecular Dissection of Complex Traits*. CRC Series, Boca Raton, FL.

Williams, C., K. Joyner, et al. 2002. Genomic consequences of interspecific *Pinus* spp. hybridisation. Biol. J. Linn. Soc. 75: 503–508.

Williams, C. and M. Reyés-Valdes. 2007. Estimating a founder's genomic proportion for each descendant in an outbred pedigree. Genome 50: 289–296

Zhou, Y., D. Gwaze, et al. 2003. No clustering for linkage map based on low-copy and undermethylated microsatellites. Genome 46: 809–816.

Chapter 4
The Female Gametophyte Inside the Ovule

Summary The sporophyte develops its specialized female reproductive structures well before meiosis can take place: the female strobilus and its megasporophylls develop integumented ovules (or megasporangia). Within each ovule's nucellar tissue forms sporogenous cells which give rise to a megaspore mother cell (MMC). The megaspore mother cell undergoes meiosis and the product of female meiosis is the linear tetrad of four megaspores. Only the megaspore closest to the chalazal end survives. The surviving megaspore divides to become a multicellular, translucent female gametophyte. Multiple egg cells, each housed in its own archegonium, will form from the female gametophyte. This haploid female gametophyte in conifers is not a synonym for an endosperm. Both gymnosperms and angiosperms have a haploid female gametophyte but only angiosperms form a triploid endosperm from multiple fertilizations. While all modern conifers follow this basic plan for female gametogenesis, the range of variation among taxa is surprising.

Each female gametophyte develops multiple egg cells, each housed in its own archegonium. This oddity has fascinated and confused plant biologists for over 150 years. Scottish naturalist Robert Brown (1773–1858), best known for his discovery of continuous motion of particles – and pollen grains – in solution (so-called Brownian motion), reported what he described as naked gymnosperm ovules in 1827 then the "plurality of embryos" in conifers and cycads in 1844. Here he saw a single *Pinus sylvestris* female gametophyte which had four to six archegonia (Brown's term: "areolae") then noted that each fertilized embryo could produce even more embryos. It is interesting to see Brown's citations of previous work: that the plurality of embryos was first discovered in cycads by French botanist M. Aubert du Petit Thouars in 1804 and that *his* discovery revived the general hypothesis about plant sexual reproduction as advanced by mathematician and inventor Samuel Morland in 1703. Following this citation thread, one can see that the discovery of gymnosperm reproduction and its peculiarities was among the enduring scientific achievements of the Enlightenment.

The scope of Brown's findings would not be fully recognized until late twentieth century paleobotanical research showed some similarities between extant

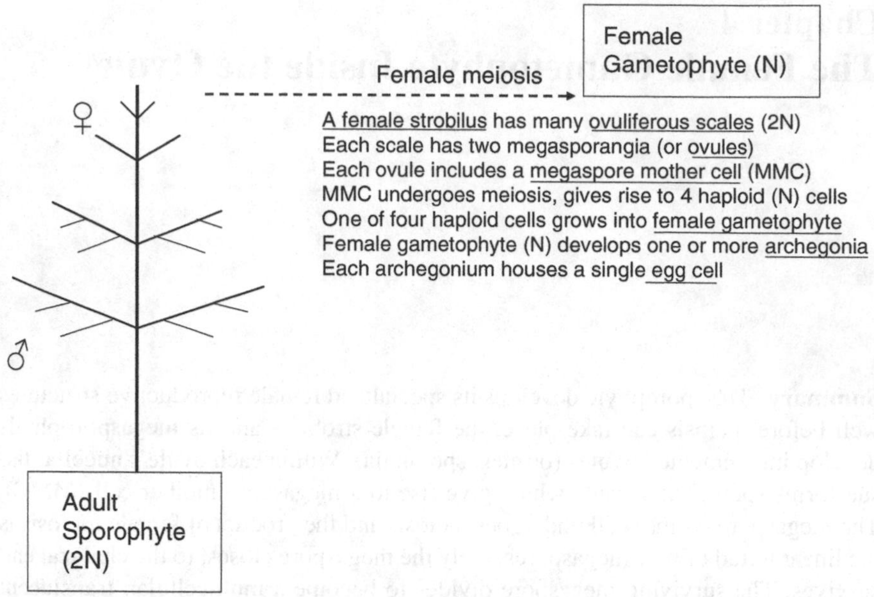

Fig. 4.1 Female strobilus development for *Pinus* spp

conifers and early seed plants. Progymnosperms and gymnosperms, both living or extinct, have a female gametophyte with multiple archegonia (Konar and Oberoi 1969; Singh 1978, p. 188; Willson and Burley 1983). Three archegonia per female gametophyte was common among the earliest non-flowering seed plants (Matten et al. 1984) and even a few angiosperm species have some form of polyembryony (Porcher and Lande 2005). Multiple archegonia are the rule, not the exception, among seed plants.

Chapter 4 starts with a description of the sporophyte's own protective tissues around the female gametophyte. The female strobilus and its cone scales support integumented ovules (or megsporangia), as shown in Fig. 4.1. A megaspore mother cell forms within the ovule then undergoes female meiosis. One meiotic product survives then develops into a endoscopic, haploid female gametophyte nested inside layers of sporophyte-derived tissues.

4.1 Female Strobilus

The adult sporophyte differentiates the female (or megasporangiate) strobilus initials within apical meristems growing on the indeterminate long shoot (Fig. 4.1). The *Pinus taeda* female strobilus initials appear by late summer (Greenwood 1980) then overwinter within the bud scales at the branch tips. Each spring, the female strobili emerge on the ends of branch tips for pollination (Table 4.1; Photo 4.1).

Table 4.1 Control-pollination protocol requires knowledge of female strobilus receptivity stages as shown here for *Pinus taeda* (Adapted from Bramlett and O'Gwynn 1980)

Stage	Megasporangiate strobilus development	Pollen application
1	One or several female strobili buds appear on vegetative shoot but they are tightly enclosed in bud scales	Too early for isolation bagging
2	Female strobili buds still enclosed inside bud scales	Isolation bagging at this stage
3	Each strobilus emerges through top of scales; they are red, pink or green in color	Too late to bag
4	Strobilus elongates, extending beyond bud scales but lower one- third or one- half of strobilus still encased by bud scales	Delay pollen application
4L	Extension from bud scales completed but ovuliferous scales have not opened fully	Pollen application possible although early
5E	Ovuliferous scales of the female strobilus are now at right angles to the cone axis	Optimum for injecting pollen into isolation bag
5L	Female strobilus remains receptive to pollen as long as the space between the ovuliferous scales is large enough for pollen to enter. These scales swell until the opening between them closes	Pollen application is late and results in low seed set
6	Female strobilus is no longer receptive because ovuliferous scales have swollen completely, closing pollen access	Applying pollen at this stage produces no seed set

In all conifers, the adult sporophyte differentiates the female (or megasporangiate) strobilus initials within apical meristems growing on the indeterminate long shoot (Fig. 4.1). The *Pinus taeda* female strobilus initials appear by late summer (Greenwood 1980) then overwinter within the bud scales at the branch tips. Each spring, the female strobili emerge on the ends of branch tips for pollination (Photo 4.1).

As a term, female strobilus refers to the stage prior to pollination, as shown in Photo 4.1 for *Pinus taeda* (Bramlett and O'Gwynn 1980). The strobilus emerges slowly from its bud then its scales flex open. Once pollinated, a female strobilus now becomes a conelet. At fertilization, the conelet becomes a cone.

The female strobilus is characterized by a series of fertile or ovuliferous scales spiraling around an axis. It is defined as a compound cone because each ovuliferous cone scale develops in the axil of a subtending bract. At the stage of full strobilus emergence, the bract will be joined to the base of the cone scale's abaxial or underside surface. A few cone scales at the base of the strobilus are sterile, lacking ovules, but the scales higher on the spiral of the axis will bear ovules.

At the base of each scale is a pair of ovules. In all cases, the ovule is exposed or naked, the defining feature for gymnosperms. The ovule's micropylar opening for the pollen faces the cone axis (Fig. 4.2). The ovule at this stage, like the cone, is derived from the adult sporophyte's diploid tissues: it is spongy nucellar tissue covered in an integument. The ovule, by definition, is unfertilized.

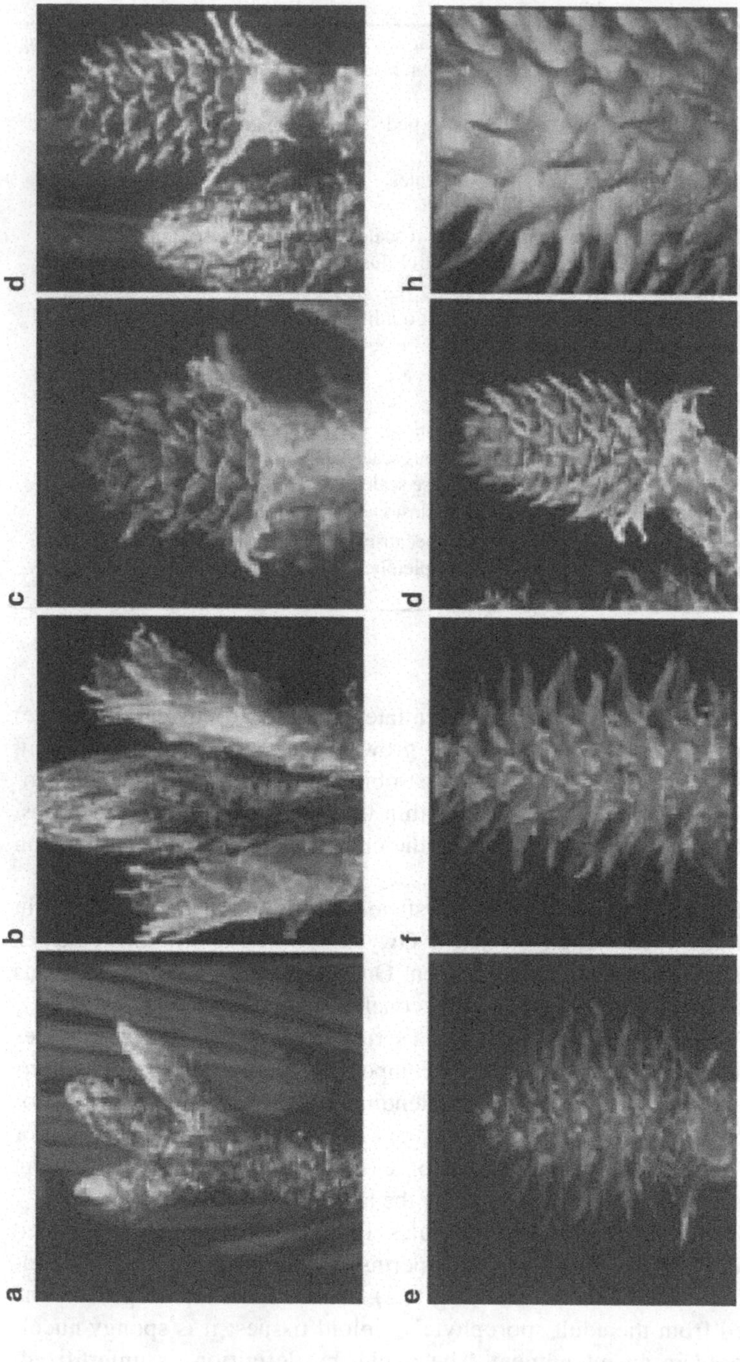

Photo 4.1a–h Stages of female strobilus receptivity for female strobilus for *Pinus taeda*. Pollination is successful when the female strobilus reaches stages 4L-5E (Photos D and E). The photographs below were used for the illustrations shown in Bramlett and O'Gwynn (1980): A is stage 2, B = stage 3, C = stage 4, D = stage 4L, E = stage 5E, F = stage 5L, G and H = Stage 6. If no pollination occurs, the strobilus dies soon after stage 5 (Photographs taken by Floyd Bridgwater, USDA-Forest Service. Permission granted)

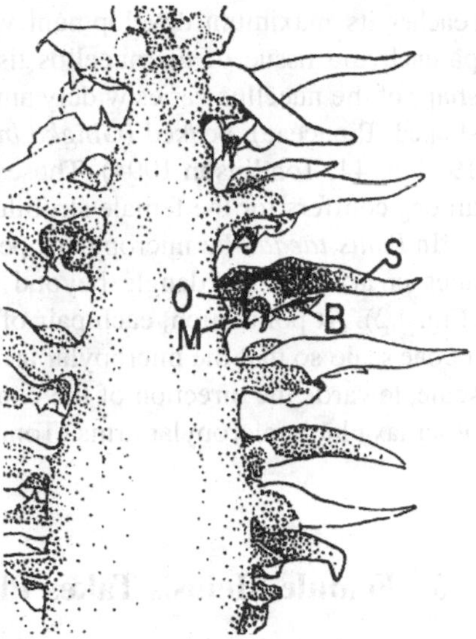

Fig. 4.2 A cross-section of the receptive *Pinus taeda* female strobilus. The ovule (O) is shown here complete with the arms of its micropyle (M) dangling downward and towards the central axis. Each cone scale (S) and its bract (B) bears a pair of ovules on its upper or adaxial surface. The pollination drop bulges out beyond the micropylar (M) arms (Drawing from Bramlett D. and C. O'Gwynn 1980. Recognizing developmental stages in southern pine flowers: the key to controlled pollinations. USDA Forest Service Southeastern Experiment Station. Permission granted)

After pollination, the ovule will house the female gametophyte enclosed in the megaspore sac. After fertilization, the ovule now becomes a developing seed; it is no longer an ovule.

For experimental purposes, it is useful to follow the receptivity of the female *Pinus taeda* strobilus. As shown in Table 4.1, stages are carefully described (Bramlett and O'Gwynn 1980) for the purposes of successful controlled pollination using isolation bags.

4.2 Ovular Anatomy

The polarity of the conifer ovule is the key to following its development. All conifers have a single integument fused around the nucellus. At one end, the integument firmly joins the nucellus at a junction defined as the chalaza. This junction is the chalazal pole or end for the ovule. The other pole is formed at the micropyle where the integument forms arms suited to pollen capture. The polarity of the chalazal-micropyle axis may be under sporophytic control but this has yet to be established.

The integument of the ovule has three distinct layers: an inner fleshy layer, middle or stony layer and an outer fleshy layer. The fate of each layer varies among gymnosperm lineages. The integument, so central to inferring the evolution of wind-pollination systems, has controversial origins (Brenner and Stevenson 2006). Late in seed development, the inner layer eventually collapses, making a papery layer within the stony layer.

The nucellus functions as the medium for pollen tube germination and, after pollination, supplies nutrients to the developing female gametophyte. The nucellus reaches its maximum development within the ovule before pollination. The soft parenchyma tissue of the nucellus tissue determines the shape of the ovule. The shape of the nucellus varies widely among conifers and other gymnosperms: dome-shaped (Pinaceae), beaked (*Gingko biloba*) or protuberant (Araucariaceae) (Singh 1978, p. 44; Tomlinson 1994). These are among the many characters which differ among conifers for the female strobilus.

In *Pinus taeda*, the micropylar opening is characterized by specialized appendages or arms which dangle beyond the nucellus and capture windborne pollen (Fig. 4.2). At pollination, each pair of ovules is inverted or oriented backwards on a cone scale so that the micropylar arms of the ovule dangle at the base of the cone scale, towards the direction of the cone axis. A few conifers have erect ovules and other taxa lack micropylar arms (Tomlinson 1994).

4.3 Female Meiosis Takes Place Inside the Nucellus

Female meiosis, described in Chapter 3, initiates the haploid phase of the life cycle. A single megaspore mother cell (MMC) deep inside the ovule's nucellus undergoes meiosis (Konar and Moitra 1980; Skinner 1992). At the completion of meiosis, the MMC gives rise to four megaspores in a linear tetrad (Fig. 4.3).

The thick, elastic megaspore wall forms around the surviving megaspore. Composed of multiple layers of exine and intine (Singh 1978, pp. 123–127; Konar and Moitra 1980), the wall is covered with sporopollenin (Konar and Moitra 1980),

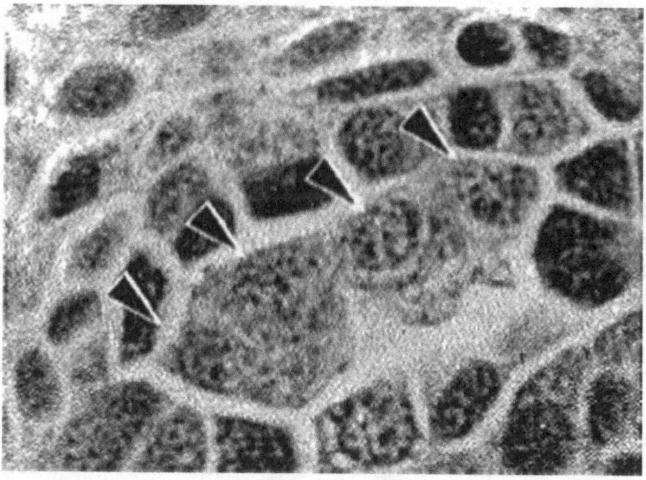

Fig. 4.3 Four haploid cells (arrowheads) form the linear megaspore tetrad after female meiosis (From D. Harrison and M. Slee. 1992. Long shoot terminal bud development and the differentiation of pollen- and seed-cone buds in *Pinus caribaea* var. *hondurensis*. Canadian Journal of Forest Research 22: 1565–1668. Copyright permission granted)

Box 4.1 Mosaicism in the female gametophyte

Rare exceptions to monospory have been reported for several taxa. In roughly 1% of *Pinus sylvestris* ovules, two megaspores from the same linear tetrad will survive (Sarvas 1962). These develop as two separate female gametophytes for a brief period then fuse so that the resulting female gametophytic tissue is a mosaic of two haplotypes. A second case is survival of all four megaspores (tetraspory) followed by fusion into a single female gametophyte. Yet another type of tetraspory occurs if one megaspore mother cell undergoes meiosis but its tetrad failed to form cell walls around the resulting megaspores. Here a single megaspore with four different nuclei forms the female gametophyte. This is the case for *Cupressus sempervirens* (El-Maataoui et al. 1998). In the final case, the megaspore mother cell does undergo meiosis and the tetrad has normal cell wall formation but two, three or four megaspores survive and together these develop into a single, fused multicellular female gametophyte for *Ginkgo biloba* and for *Pinus* spp. (O'Malley et al. 1988; O'Malley and Kelly 1988). In all cases, the female gametophyte develops as a genetic mosaic composed of multiple maternal haplotypes as opposed to monospory where the female gametophyte develops from a single megaspore and thus has a single DNA haplotype.

the wall protects the surviving megaspore as it develops into the female gametophyte. Although the megaspore wall is derived from adult sporophyte tissues, it also has a basal layer composed of haploid female gametophyte cells (Fig. 4.4). Unlike the spore wall for pollen grains, no tapetal cells surround the megaspore wall (Singh 1978, p. 127).

Monospory is the general rule in conifers and other gymnosperms (Willson and Burley 1983; Lill 1976) but its exceptions are notable and unusual (Box 4.1). Normally, only one of the four megaspores survives and this will be the megaspore at the chalazal end (upper right corner in Fig. 4.3). This means that the assumption of monospory should be validated. If monospory is correctly assumed then the female gametophyte's DNA haplotype can be used as a proxy for the egg cell's own haplotype.

4.4 From Megaspore and Monospory to Haploid Female Gametophyte

One of the five critical features for the diplohaplontic life cycle of seed plants is retention of the female gametophyte within the sporophyte (Fig. 4.4). The surviving megaspore is slow to develop into a female gametophyte. It moves through several well-defined stages: the free nuclei stage, the cellularization stage and the cellular

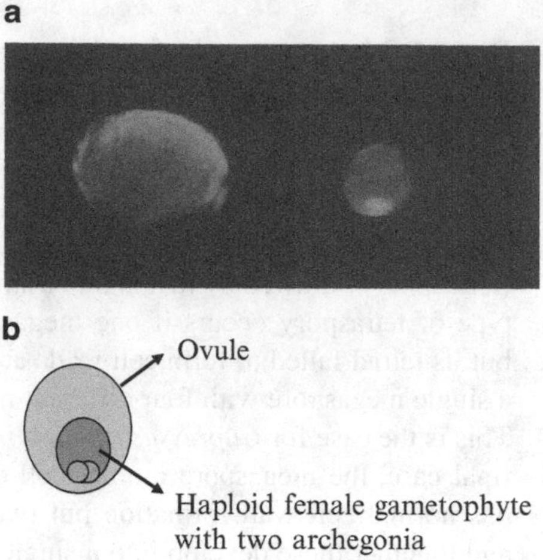

Fig. 4.4 An intact *Pinus taeda* ovule just prior to fertilization (**a**). Beside the ovule is its female gametophyte. Archegonia within the female gametophyte transmit the brighter fluorescein diacetate dye signal. A schematic diagram of the ovule components are shown in (**b**) (Photographs from Williams 2008. With permission)

growth stages, all of which have been documented for many conifers and gymnosperms (Konar and Oberoi 1969; Konar and Moitra 1980; Skinner 1992).

4.4.1 Free Nuclei Stage

By May, the functional megaspore has enlarged until it is many times the size of all other cells in the ovule (Skinner 1992). The enlarging megaspore has also divided mitotically to form the free nuclear gametophyte (Fig. 4.5). Nuclei number reaches 2,500 for *Pinus roxburghii* (Konar and Oberoi 1969). A large vacuole develops in the center and the nuclei cluster at its periphery. The free nuclear gametophyte is transformed into the cellular phase via alveolus formation (Konar and Moitra 1980) although the process itself is not entirely clearcut and given to misinterpretation (Singh 1978, p. 114).

4.4.2 Cellularization Stage

The female gametophyte looks like a honeycomb. Each cavity in the honeycomb is an alveolus. Within each alveolus, spindles form around a nucleus and appear to lay down the cell wall materials around each of the free nuclei (Konar and Moitra 1980; Singh 1978, p. 113). In *Pinus taeda*, cellularization is also the stage where a few cells at the micropylar end become the archegonial initials (Skinner 1992). Each initial forms a central cell and a small neck of the vase-shaped archegonium; the neck cells will facilitate the passage of the pollen tube (Konar and Moitra 1980). The central cell enlarges and becomes highly vacuolated (the foam stage). The vacuolated archegonium divides into a ventral canal cell and an egg cell. The egg cell increases in size before the jacket layer forms around it.

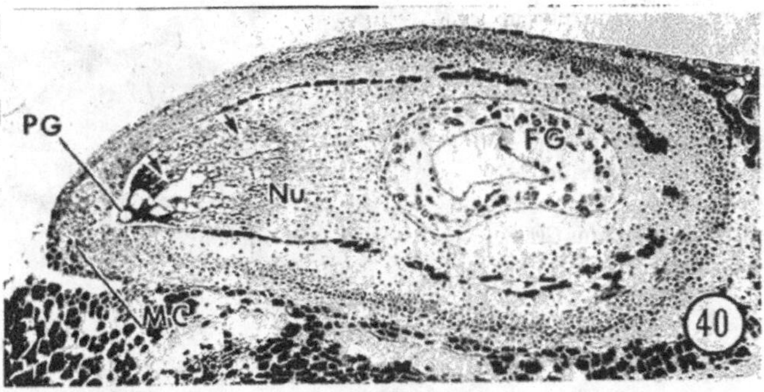

Fig. 4.5 Free nuclear stage of the female gametophyte (FG) after pollination. The micropylar canal (MC) is occluded and the pollen tubes (arrows) from the pollen grains (PG) have grown through the nucellus (Nu) towards the free-nuclear female gametophyte (FG) (From Harrison D. and M. Slee. 1992. Long shoot terminal bud development and the differentiation of pollen- and seed-cone buds in *Pinus caribaea* var. *hondurensis*. Canadian Journal of Forest Research 22: 1565–1668. Permission granted)

4.4.3 Cellular Growth Stage

Differentation now occurs within the female gametophyte and this occurs well after pollination. In particular, the *Pinus taeda* female gametophyte differentiates a few egg cells near its micropylar end and each egg cell is housed in an archegonium (Photo 4.2). In the Pinaceae, embedded archegonia are clustered at the micropylar end of the female gametophyte. Archegonia swell near the tent pole at the distal end. By fertilization, the female gametophyte is shaped like a rice grain and surrounded by nucellar tissue.

The gametophytic cells around each archegonium are transformed into a special layered covering, the archegonial jacket, by late May. Conifers and gymnosperms differ widely as to the presence and morphology of the archegonial jacket (Konar and Moitra 1980) but the thick jacket wall is pitted in *Pinus* spp. and it has been proposed that it is through these pits that the archegonium maintains transport of macromolecules (Konor and Moitra 1980). The female gametophyte continues to enlarge until it is time for fertilization.

4.5 Variations in Female Gametophyte Development

In *Pinus strobus*, the female gametophyte overwinters in the free nuclear phase as a tiny spherical sac with 32 free nuclei embedded in cytoplasm (Ferguson 1904). The megaspore wall forms the sac. In spring, roughly 2,500 free nuclei are added to cytoplasm so that the free nuclei resemble “a finely granulated matrix very much like the suspension of fruit in a gelatine dessert” (Emig 1935). The number of free

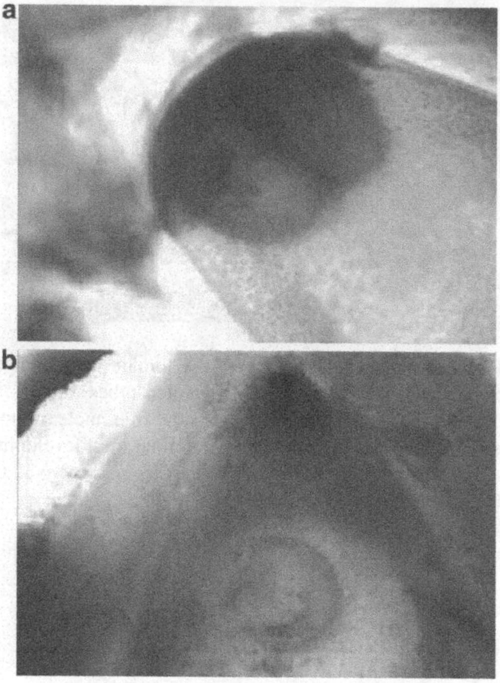

Photo 4.2 The megaspore membrane has been cut away from the haploid female gametophyte at the micropylar end to show a *Pinus taeda* ovule with two archegonia prior to fertilization (**a**) and another *Pinus taeda* ovule with a single archegonium (**b**) (Photographs from Williams 2008. With permission)

nuclei varies widely among other taxa, ranging from 256 in *Taxus baccata* to 8,000 in *Gingko biloba* (Konar and Moitra 1980).

The number of archegonia varies widely by taxa (Table 4.2). For nearly all gymnosperms, whether monoecious or dioecious, multiple archegonia has long been the rule rather than the exception (Konar and Oberoi 1969; Singh 1978, p. 188; Willson and Burley 1983) since multiple archegonia were first reported more than 150 years ago for *Pinus sylvestris* (Brown 1844). Note that the largest number of archegonia can be counted among the Podocarpaceae. These have as many as 25 archegonia per female gametophyte (Konar and Oberoi 1969; Willson and Burley 1983). Multiple archegonia of the Cupressaceae are grouped into complexes at the micropylar end so that archegonial numbers tend to be higher. Chalazal archaegonia, although rare among conifers, are typical of *Cedrus deodara* (Pinaceae) and a few other taxa (Konar and Moitra 1980).

4.6 Ploidy Levels for the Female Gametophyte in Conifers

For mature female gametophytes, the nuclei ploidy levels are highly uniform and haploid (1C) within the Pinaceae family; *Abies*, *Cedrus* and *Pinus* spp. all had highly uniform ploidy levels for female gametophytes relative to the diploid DNA

Table 4.2 Archegonial numbers by taxa, with or without complexes

Taxon	Archegonial number	Reference
Librocedrus decurrens	10–15	Lawson 1907
Thuja plicata	7–9	Owens and Molder 1980
Thuja orientalis	5–8	Singh and Oberoi 1962
Sequoia sempervirens	12–17	Lawson 1904
		Buchholz 1939
Metasequoia	10	Engles and Gianordoli 1983
Tsuga heterophylla	3; 2–5	Stanlake and Owens 1974
Abies grandis	2–3	Singh and Owens 1982
Larix occidentalis	2–5	Owens and Molder 1979
Picea glauca	3; 1–4	Runions and Owens 1999
Pinus sylvestris	1–3	Sarvas 1962
Pinus taeda	1–4	Skinner 1992
Pinus radiata	2; 1–4	Lill 1976
Pinus strobus	1–5	Ferguson 1904
Pinus contorta	2–4	Owens et al. 1982
Pinus virginiana	2–7	Thomas 1951
Pinus monophylla	3	Haupt 1941
Pinus monticola	3–5	Owens and Molder 1977
Pinus lambertiana	5	Haupt 1941

content of the embryo (Pichot and El-Maataoui 1997). But this was not the case for genera sampled within the Cupressaceae. Ploidy levels for female gametophyte tissue ranges from 1C to 6C, a total of six different ploidy levels although the embryos developing inside the same female gametophytes were uniformly diploid. For this conifer family, ploidy levels appear to be relaxed in female gametophyte tissue yet this variability in ploidy is not transmitted to either the archegonial tissues (egg cell) or the fertilized zygote (Pichot and El-Maataoui 1997).

Again, conifers have a haploid female gametophyte but this gametophyte is not synonymous or even analogous to an endosperm. Both gymnosperm and angiosperm plants have a female gametophyte but only angiosperm plants have an endosperm.

4.7 Variants in Microspore and Pollen Morphology for Conifers

The *Pinus taeda* female strobilus (Photo 4.1a–h) presents only one of many variants for strobilus morphology (Table 4.3). Some conifer taxa have megasporangiate strobili in terminal positions on the branch while others have strobili in axillary positions (Table 4.3). Strobili can be simple or compound. Bract length and shape varies widely. Among the Pinaceae, *Pseudotsuga* has long bracts but *Pinus* has

Table 4.3 A list of contrasting characters for female strobilus morphology between the Pinaceae and other conifers or gymnosperms. The character list was adapted from Hart (1987)

Character	Taxon/taxa	Character	Contrast	Taxon/taxa
Ovulate strobilus				
Position	*Pinus*	Terminal	Axillary	*Cupressus* (Cupressaceae)
Number	Pinaceae	Simple	Compound	*Taxus* (Taxaceae)
Bract shape	Pinaceae	Free	Fused	*Phyllocladus* (Podocarpaceae)
Cone orientation	*Pinus*	Pendulous	Upright	*Abies*, *Cedrus*, *Keteleeria*, *Pseudolarix* (Pinaceae)
Female gametophyte				
Pollination drop	*Pinus*, *Picea*	Present	Absent	*Abies*, *Larix* (Pinaceae)
Pollen germination	Pinaceae	On nucellus	On scale	Araucariaceae
Shape of micropyle	*Pinus*	Symmetrical	Asymmetrical	*Larix*, *Pseudotsuga* (Pinaceae)
Archegonial configuration	Pinaceae	Separate	Grouped in complexes	Cupressaceae
Archegonial position	Pinaceae	Micropylar end	Middle	*Sequoia*, *Sequoidendron* (Cupressaceae)
Archegonial jacket	Pinaceae	Present	Absent	*Sequoia*, *Sequoidendron* (Cupressaceae)
Suspensor anchorage	Pinaceae	Outside archegonium	Inside archegonium	Araucariaceae
Proembryo	*Pinus*	Fills archegonium	Incomplete filling	*Fitzroya* (Cupressaceae)
Cleavage embryony	*Pinus*	Present	Absent	*Picea* (Pinaceae)
Ovules or seeds				
Position	*Pinus*	Inverted ovules	Erect	*Taxodium* (Cupressaceae)
Number	*Pinus*	Two ovules per scale	One ovule per scale	*Juniperus* (Cupressaceae)
Presence of aril	*Pinus*	No aril	Aril	*Taxus*
Seed morphology	*Pinus*	Winged	Unwinged	*Gingko biloba* (Ginkgoaceae)

short bracts and members of the Podocarpaceae have fused bracts (Table 4.3). Other than *Pinus*, many genera within the Pinaceae have pendulous female strobili (Table 4.3). Each cone scale bears either one or two ovules on its upper or adaxial surface (Table 4.3). Conifers have a surprising degree of variation in female reproductive development.

4.8 Closing

Robert Brown's discovery of the female gametophyte and its multiple archegonia was more profound than he knew because it is a condition inherent to early seed plant lineages, progymnosperm and gymnosperm. The female gametophyte and its multiple archegonia are one of the most highly conserved features for seed plant reproduction.

Beyond this *Bauplan* feature, female reproductive structures vary widely among conifers and other gymnosperms. Here too heterospory opened new selection space for the divergence of female and male sporogenesis but also for sporophyte's own modifications for female and male strobili. Variation among sporophyte-derived tissues are far greater than those found in the gametophyte phases.

References

Bramlett, D. and C. O'Gwynn. 1980. Recognizing developmental stages in southern pine flowers: the key to controlled pollinations, USDA Forest Service Southeastern Experiment Station, 14 pages.

Brenner, E. and D. Stevenson. 2006. Using genomics to study evolutionary origins of seeds. Editor: C.G. Williams. In: *Landscapes, Genomics and Transgenic Conifers*. Springer, Dordrecht, The Netherlands.

Brown, R. 1844. On the plurality and development of the embryos in the seeds of Coniferae. Annual Magazine of Natural History 13: 368–374.

Buchholz, J. 1939. The embrogeny of *Sequoia sempervirens*. American Journal of Botany 26: 248–257.

El-Maataoui, M., C. Pichot, et al. 1998. Cytological basis for a tetraspory in *Cupressus sempervirens* L. megagametogenesis and its implications in gentic studies. Theoretical Applied Genetics. 96: 776–779.

Emig, W. 1935. The megagametophyte of *Pinus*. I. Introduction. American Journal of Botany 22: 500–503.

Engels, F. and M. Gianordoli. 1983. The basic anatomy of *Metasequoia* female gametophytes. Acta Botanica Neerlandica 32: 295–305.

Ferguson, M. 1904. Contributions to the life history of *Pinus* with special reference to sporogenesis, the development of gametophytes and fertilization. Proceedings of the Washington Academy of Sciences 6: 1–202.

Greenwood, M. 1980. Reproductive development in loblolly pine. I. The early development of male and female strobili in relation to the long shoot growth behavior. American Journal of Botany 67: 1414–1422.

Haig, D. 1992. Brood reduction in gymnosperms. Editors: M. Elgar and B. Crespi. In: *Cannibalism: Ecology and Evolution Among Diverse Taxa*. Oxford University Press, Oxford, pp. 62–84.

Harrison, D. and M. Slee. 1992. Long shoot terminal bud development and the differentiation of pollen- and seed-cone buds in *Pinus caribaea* var. *hondurensis*. Canadian Journal of Forest Research 22: 1565–1668.

Hart, J. 1987. A cladistic analysis of conifers: preliminary results. Journal of Arnold Arboretum 68: 269–307.

Haupt, A. 1941. Oogenesis and fertilization in *Pinus lambertiana* and *P. monophylla*. Botanical Gazette 102: 482–498.

Konar, R. and A. Moitra. 1980. Ultrastructure, cyto- and histochemistry of female gametophyte of gymnosperms. Gamete Research 3: 67–97.

Konar, R. and Y. Oberoi. 1969. Recent work on reproductive structures of living conifers and taxads – a review. Botanical Review 35: 89–116.

Lawson, A. 1904. The gametophytes, archegonia, fertilization and the embryo of *Sequoia sempervirens*. Annals of Botany 18: 1–28.

Lawson, A. 1907. The gametophytes and embryo of the Cupressineae with special reference to *Libocedrus decurrens*. Annals of Botany 21: 281–301.

Lill, B. 1976. Ovule and seed development in *Pinus radiata*, postmeiotic development, fertilization and embryogeny. Canadian Journal of Botany 54: 2141–2154.

Matten, L., T. Fine, et al. 1984. The megagametophyte of *Hydrasperma tenuis* long from the uppermost Devonian of Ireland. American Journal of Botany 71: 1461–1464.

O'Malley, D. and J. Kelly. 1988. Genetic analysis of a megagametophyte color polymorphism in *Gingko biloba*. Journal of Heredity 79: 51–53.

O'Malley, D., R. Guries, et al. 1988. Electrophoretic evidence for mosaic 'diploids' in megagametophytes of knobcone pine (*Pinus attentuata* Lemm.). Silvae Genetica 37: 85–88.

Owens, J. and M. Molder. 1977. Seed-cone differentiation and sexual reproduction in western white pine (*Pinus monticola*). Canadian Journal of Botany 55: 2574–2590.

Owens, J. and M. Molder. 1979. Sexual reproduction of *Larix occidentalis*. Canadian Journal of Botany 57: 152–169.

Owens, J. and M. Molder. 1980. Sexual reproduction in western red cedar (*Thuja plicata*). Canadian Journal of Botany 58: 1376–1393.

Owens, J., S. Simpson, et al. 1982. Sexual reproduction of *Pinus contorta*. II. Postdormancy ovule, embryo and seed development. Canadian Journal of Botany 60: 2071–2083.

Pichot, C. and M. El-Maataoui. 1997. Flow cytometric evidence for multiple ploidy levels in the endosperm of some gymnosperm species. Theoretical and Applied Genetics 94: 865–870.

Porcher E and Lande R. 2005. Reproductive compensation in the evolution of plant mating systems. New Phytologist 166: 673–684.

Runions, C. and J. Owens. 1999. Pollination of *Picea orientalis* (Pinaceae): saccus morphology governs pollen buoyancy. American Journal of Botany 86: 190–197.

Sarvas, R. 1962. Investigations on the flowering and seed crop of *Pinus silvestris*. Communicationes Instituti Forestalis Fennica 53: 1–198.

Singh, H. 1978. *Embryology of Gymnosperms*. Gebruder Borntraeger, Berlin.

Singh, H. and Y. Oberoi. 1962. A contribution to the life history of *Biota orientalis* Endl. Phytomorphology 12: 373–393.

Singh, H. and J. Owens. 1982. Sexual reproduction in *Abies grandis*. Canadian Journal of Botany 60: 2197–2214.

Skinner, D. 1992. Ovule and embryo development, seed production and germination in orchard grown control pollinated loblolly pine (*Pinus taeda* L.) from coastal South Carolina. Department of Biology. University of Victoria, Victoria, BC, 88 pp.

Stanlake, E. and J. Owens. 1974. Female gametophyte and embryo development in western hemlock (*Tsuga heterophylla*). Canadian Journal of Botany 52: 885–893.

Thomas, R. 1951. Reproduction in *Pinus virginiana* Miller. Vanderbilt University, Nashville TN, 80 p.

Tomlinson, P. 1994. Functional morphology of saccate pollen in conifers with special reference to the Podocarpaceae. International Journal of Plant Sciences 155: 699–715.

Williams, C. 2008. Selfed embryo death in *Pinus taeda*: a phenotypic profile. New Phytologist 178: 210–222.

Willson, M. and N. Burley. 1983. *Mate Choice in Plants*. Princeton University Press, Princeton, NJ.

Chapter 5
The Male Gametophyte Enclosed in a Pollen Wall

Summary All modern conifers follow a basic plan for male reproduction: the male strobilus is composed of fertile cones scales or sporophylls attached to a central cone axis. Each sporophyll has microsporangia or pollen sacs attached to its underside. Each pollen sac contains many pollen mother cells (PMC) and each pollen mother cell undergoes meiosis then gives rise to four microspores after meiosis. Each microspore develops into a multicellular, mobile male gametophyte enclosed inside a pollen wall. Pollen grains are released for aerial transport by dehiscence of the male strobilus. Although most pollen grains do fall near the adult tree, a small fraction will travel hundreds of kilometers from source.

Each spring, allergy sufferers seize upon the yellow coating of pine pollen on car windshields as the source of their suffering. The reality is that *Pinus* spp. pollen rarely triggers allergies yet U.S. medical professionals are quoted as saying that pine pollen does not cause allergies because "pine pollen is too heavy and therefore is not widely disseminated".[1] Nothing could be farther from the truth: pine pollen can travel for hundreds or even thousands of kilometers from source. Biomedical interests aside, the more interesting question is how far does pine pollen move?

To this end, the most enduring answer comes from Professor Gunnar Erdtman's Atlantic Ocean pollen experiments. On the eve of World War II, this eminent pollen biologist boarded the M.S. Drottningholm, a seafaring passenger ship from Gothenburg Sweden on May 29. The journey, by design, coincided with conifer pollen release at northerly latitudes, namely Sweden, Finland and Canada because his goal was to sample forest tree pollen using two Hoover Electrolux vacuum cleaners. He had already established that each horizontal metal cylinder could suction airborne pollen for hours on end, capturing as little as one pollen grain per 1,000 m^3 of air. To each vacuum cleaner he had added a filter and a manometer in place of the dust bag. Once on board, one vacuum was placed in the ship's rigging at 18 m above sea level and the other on top of the ship's bridge.

Using the vacuum method and a microscope in his cabin, Professor Erdtman sampled pollen at all transects of the ship's route before his arrival in New York.

[1] http://www.aaaai.org/aadmc/ate/ate/category.asp?cat=10171, Accessed November 9, 2008.

Table 5.1 Pollen collection data reported by Erdtman (1937) from May 29 to June 7 1937. Note that the vacuum cleaners ran the longest in Section V but that the largest pollen count came from other samples

Sampling		Location	*Pinus* spp. pollen grain count
Section I	May 30	North Ssea	157
Section II	May 31	Midway between Ireland–Iceland	64
Section III	June 1	Mid-ocean > 1,000 km from land	21
Section IV	June 2	650 km from land	23
Section V	June 4	250–660 km off Newfoundland	28
Section VI	June 5	300 km south of Newfoundland	66
Section VII	June 6	220–300 km off Nova Scotia and Massachusetts coasts	14

Two days later he boarded the same ship, now returning to Sweden. On this return trip he collected no pollen (Erdtman 1937). His pollen sampling data are shown in Table 5.1. From Section III in this experiment, one can see that *Pinus* spp. pollen moved 1,000 km from land.

Like Chapter 4, the story of male reproductive development (Fig. 5.1) in conifers starts with the adult sporophyte which protects the male gametophyte even after its release from the sporangial sac. The male strobilus is composed of many microsporophylls attached to a central axis. Each microsporphyll has sporangial sacs on its abaxial surface. Microsporangial cells divide in several planes, creating a mass of sporogenous cells, many of which will undergo meiosis. The product of each male meiosis will be a tetrad of four microspores which break apart. Each single-celled microspore will eventually be protected by a spore wall as it divides into a multicellular male gametophyte. A pollen grain is the male gametophyte enclosed in a spore wall secreted by sporophyte-derived tapetal cells.

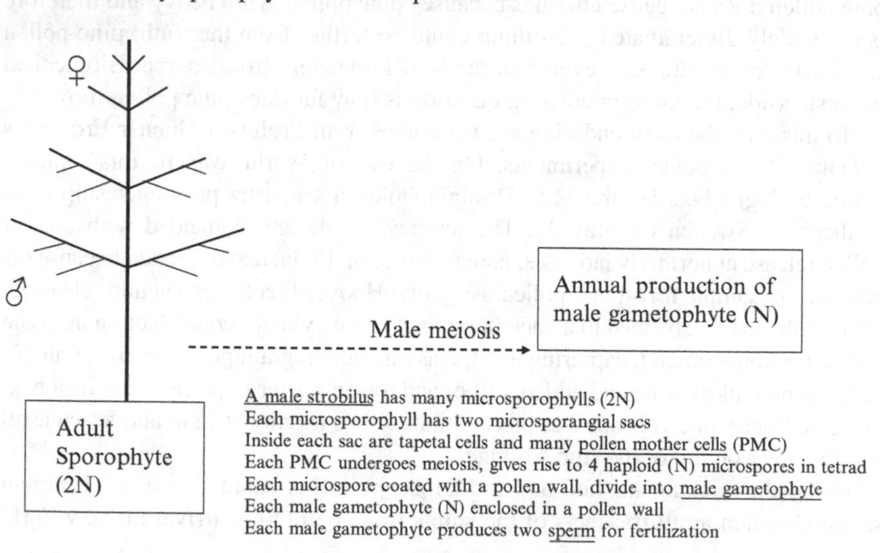

Fig. 5.1 Male strobilus development for *Pinus taeda*

The spore wall (or pollen wall or exine) protects the pollen grain from early development in the sporangial sac to flight to pollen capture and finally, its germination.

5.1 Male Strobilus

A closer look at a single male strobilus (also called microsporangiate strobilus or pollen cone) shows many flexible scales (Photo 5.1). Each scale is a microsporophyll and each microsporophyll supports two microsporangial sacs (Fig. 5.2). The sporogenous tissues inside each sac will form pollen mother cells (PMC) (Fig. 5.3).

Photo 5.1 A cluster of immature *Pinus torreyana* male strobili (Photograph taken by the author)

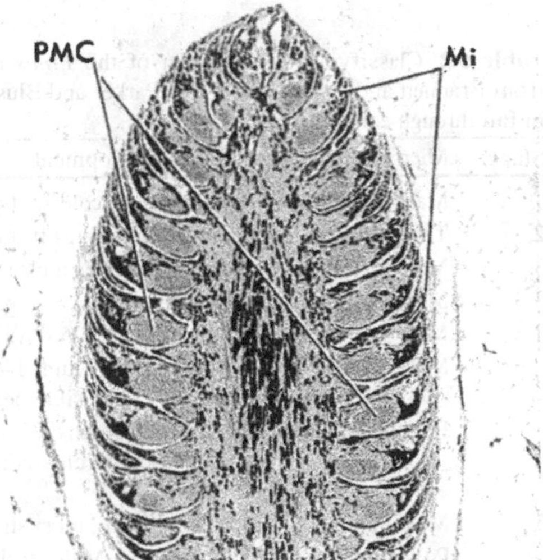

Fig. 5.2 A median section of a male strobilus (or pollen cone) after the completion of microsporophyll (Mi) initiation of *Pinus caribaea.* (From D. Harrison and M. Slee, 1992. Long shoot terminal bud development and the differentiation of pollen- and seed-cone buds in *Pinus caribaea* var. *hondurensis.* Canadian Journal Forest Research 22: 1565–1668. Copyright permission granted)

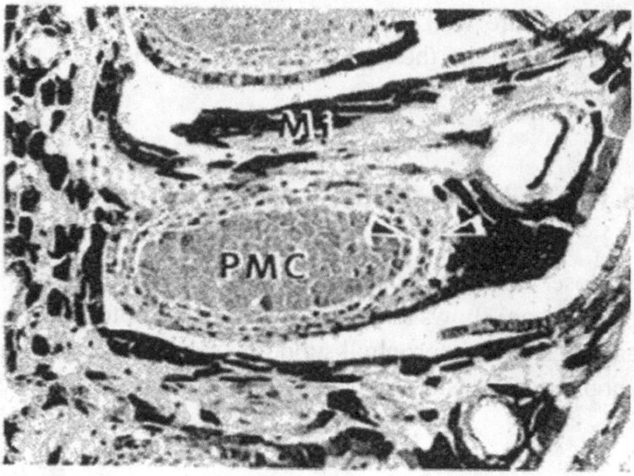

Fig. 5.3 A longitudinal section of a microsporophyll (Mi) shows the pollen mother cells (PMC) within a microsporangium. The pollen mother cells are surrounded by two layers of tapetal cells. (From D. Harrison and M. Slee. 1992. Long shoot terminal bud development and the differentiation of pollen- and seed-cone buds in *Pinus caribaea* var. *hondurensis*. Canadian Journal Forest Research 22: 1565–1668. Copyright permission granted)

Pinus taeda pollen development has its own classification system (Bramlett and Bridgwater 1989; Table 5.2). Stage 1 occurs in the autumn when tiny male strobili can be seen enclosed in bud scales on the vegetative shoot. Stage 2 marks the emergence of the individual male strobilus from the bud scales; the male strobilus slowly elongates from November to February in Northern Hemisphere countries. Stage 3 marks increased length of the strobili during which the strobili exude a clear liquid when pressed.

The *Pinus* spp. male strobilus is generally green (Photo 5.1) then yellow and finally golden brown upon dehiscence. As the male strobilus turns from green to yellow, it can be characterized by liquid exudation. In Stage 3.3, the male strobilus exudes yellow fluid when pressed. At Stage 3.6, the male strobilus exudes clear liquid and this occurs 3–5 days before release. At Stage 3.9, the strobilus is now

Table 5.2 Classifying development of the *Pinus taeda* microsporangiate strobilus (Modified from Bramlett and Bridgwater 1989; Parker and Blush 1996; Williams 2008). Stages 1 to 3 occur in fall through early spring

Stage	Microsporangiate strobilus development
1	Male strobili are encased in bud scales at tips of vegetative shoot
2	The individual male strobilus emerges from its bud scales
3	Male strobilus lengthens and exudes a clear liquid when pressed
3.3	Male strobilus exudes yellow fluid
3.6	Male strobilus exudes clear fluid; 3–5 days from dehiscence
3.9	Male strobilus exudes little, if any fluid; 1–2 days from dehiscence. Microsporophylls bend easily so that spaces are visible between them
4	Proximal end of male strobilus begins releasing pollen and dehiscence moves acropetally; less than 10% of pollen released; pollen release takes 7–14 days between stages 4 and 6
5	Maximum stage for pollen release; most strobili within a cluster are dehiscent
6	Pollen release is now completed. Male strobilus becomes dry and brown in color

1–2 days from dehiscence and pollen release. Little fluid, if any, is exuded when pressed. At this stage, the strobilus flexs easily. Spaces between microsporophylls are now visible (Table 5.2).

At Stage 4, pollen release begins at the proximal end of the male strobili then progresses acropetally. At this stage, less than 10% of the pollen is released. Stage 5 marks maximum release for a cluster and here the majority of male strobili in each cluster are releasing pollen. Stage 6 marks complete pollen release; strobili are dry, lightweight and brown in color. Pollen shed at stages 4–6 takes roughly 7–14 days (Bramlett and Bridgwater 1989; Parker and Blush 1996).

5.2 Sporangial Sacs Attached to Each Microsporophyll

Each sac contains sporogenous cells which divide into primary parietal cells and pollen mother cells (Singh 1978, pp. 6–11). The primary parietal cells divide further to form a microsporangial wall but the innermost cells function as the tapetum. The tapetal cells, interconnected to one another by cytoplasmic channels, nourish the sporogenous cells and, after meiosis, secrete some or all of the pollen wall around each microspore. The primary sporogenous cells divide into sporogenous cells which become pollen mother cells (also known as microspore mother cells), as shown in Fig. 5.3.

5.3 Microspore Polarity Determined During Male Meiosis

After male meiosis, each pollen mother cell splits into a rounded tetrad of four haploid nuclei before cell wall formation. Unlike female meiosis where only one of the four survives, all four microspores will develop into male gametophytes. Male meiosis in *Pinus* spp. is not synchronous among sporangial sacs so the stage of meiosis varies among sacs within the same male strobilus (Singh 1978, p. 25).

Tetrad formation confers each microspore's orientation or polarity over the course of pollen development (Rudall and Bateman 2007). Each of the four microspores has a proximal and a distal pole. The proximal pole points towards the center of the tetrad (so that it is inward facing) and its distal pole points away from the tetrad's center (outward facing).

As soon as the tetrad forms, each microspore will form two air-filled bladders or sacci on its distal face. The aperture (sometimes referred to as the sulcus, leptoma or germinal pore) is distal, located between the two sacci. The pollen tube will emerge from the distal aperture. The aperture and its pollen tube are the most highly conserved pollen characters in seed plants (Pettit 1985).

5.4 Male Gametophyte Enclosed in Pollen Wall

The two-layer pollen wall is composed of a rigid exine and a flexible intine. While the sporophyte's diploid tapetal cells clearly deposit the exine during and after the tetrad stage (Dickinson and Bell 1976), the origin of the intine could be either sporophytic or gametophytic. Note that the following description of *Pinus* spp. is distinct from pollen wall zonation for *Cupressus* spp. (Chichiricco and Pacini 2008).

5.4.1 *Exine Formation*

At the end of male meiosis, the rounded tetrad holds its four microspores within a callose wall. Tapetal cells around the tetrad, interconnected by broad cytoplasmic channels, now secrete the exine for each microspore. These secretions are sporopollenin-rich (Dickinson and Bell 1976). At first, each microspore inside the tetrad is sharply angular then it slowly rounds out as the exine layer is formed (Harrison and Slee 1992).

The exine is composed of the two major layers: the outer sexine and the inner nexine. The inner nexine is further composed of nexine I and nexine II. The sexine and nexine I are both laid down when microspores are still enclosed within the tetrad's callose wall but these two layers do not cover the same parts of the pollen grain. The sexine layer covers the area where sacci will form and the central capsule but not at the distal aperture. The nexine I layer covers the central capsule but not the sacci (Rowley et al. 2000).

The sexine and nexine layers differ with respect to diffusion: the sexine around the sacci is highly permeable for molecules up to 200 nm in diameter (Bohne et al. 2003). The nexine is the opposite: it acts as an ultrafilter membrane, blocking most proteins and allowing only molecules up to 4 nm in diameter (Bohne et al. 2003). The functional importance of the sacci's rapid polymer exchange and the central capsule's slow exchange may have functional importance during pollination (Bohne et al. 2005).

Now each microspore, now rounded with two sacci, breaks free from the tetrad's callose wall. The tapetal cells continue to deposit sporopollenin on the sexine layer but now the nexine II layer starts to forms. The nexine II layer covers only the body of the pollen grain (also called the central capsule) such that the sacci are covered only by the sexine layer. In the final stages, the sacci will lack not only the nexine layer but also the last layer of the pollen wall, the intine (Dickinson and Bell 1976).

5.4.2 *Intine Formation*

The intine is a thin inner layer of the pollen wall rich in hemicelluloses. The hemicellulose content is more flexible than the rigid sporopollenin-rich layers of the exine.

The intine is composed of two layers, the outer intine and the inner intine. These two layers differ with respect to what is covered and when the covering is added. The outer intine forms before the first mitosis of the microspore and its coverage is incomplete. The outer intine covers the central capsule but not the distal aperture between the sacci. Now the microspore proceeds through two mitoses before the inner intine appears. This occurs after the formation of the two prothallial cells (Rowley et al. 2000). The inner intine forms a continuous covering around both the central capsule and the distal aperture. Only the inner intine will cover the pollen tube at germination (Pettit 1985).

5.5 From Microspore to Male Gametophyte

By pollen wall completion, the young microspore has already started its series of asymmetric mitoses (Fig. 5.4). Each mitosis gives rise to two cells, one large and one small. Only the large cell will continue through the mitotic progression.

The asymmetric mitoses of the male gametophyte is a curious yet universal feature of conifers and other gymnosperms (Rudall and Bateman 2007) but they signal the development of specialized cell types (Fig. 5.4). For *Pinus* spp., pollen grains halt at the four-cell stage at release from the sporangial sacs; this will be the generation (or body) cell and the stalk cell (Fig. 5.4). The final mitotic division takes place just prior to fertilization when the body (or generative) cell divides into two sperm nuclei.

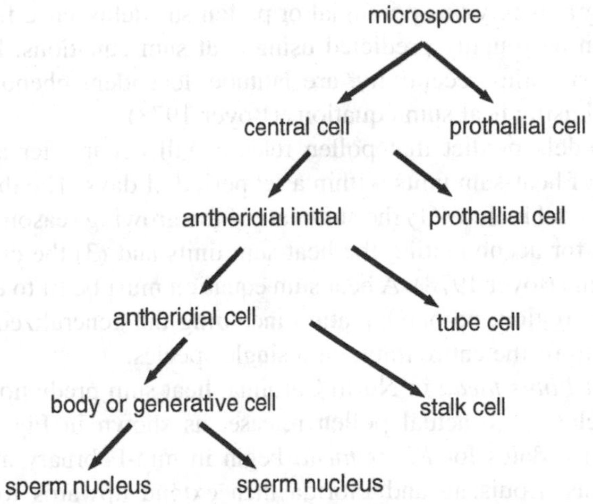

Fig. 5.4 The single-celled microspore divides via a series of asymmetric mitoses to become a multicellular male gametophyte. After mitosis, only the larger cells (listed under the microspore) divides again; the smaller cell does not divide further. Terminology adheres to Singh (1978)

Box 5.1 Released pine pollen lacks pre-packaged proteins

Mature pollen cells contain mRNA transcripts needed for germination and early pollen tube elongation but oddly, protein synthesis in *Pinus* spp. halts starting at pollen release and does not resume again until pollen germination (Pettit 1985; Frankis 1990). This is one of the more intriguing features distinguishing conifer pollen from angiosperm pollen (Frankis 1990; Fernando et al. 2001) but its adaptive significance is not fully understood. This peculiar feature opens the question of how a pollen grain responds to abiotic stress if it cannot modify its protein complement during flight. At early stages of germination, the generative cell and tube cell nuclei actively synthesize additional RNA; most of these transcripts are thought to be ribosomal RNA (Young and Stanley 1963; Frankis 1990).

The question of RNA transcript partitioning among cells within a pollen grain has yet to be addressed for conifers or gymnosperms. Transcript partitioning refers to the process by which mRNA transcripts destined for sperm nuclei are synthesized at the microspore level then progressively partitioned until they are localized in one or more cells (Engel et al. 2003). Comparative localization and identification of pollen mRNA transcripts between gymnosperm and angiosperm taxa is the next logical research step.

5.6 Predicting Time of Pollen Grain Release

Conifer pollen grains rely on sporangial or pollen sac dehiscence for release. This drying event can be roughly predicted using heat sum equations. Pollen shed, as well as female strobilus receptivity, are latitude-dependent phenomenona which can be predicted using heat sum equations (Boyer 1978).

Heat-sum models predict that pollen release will occur after accumulating a critical number of heat-sum units within a set period of days. The three parameters for a heat sum model include (1) the start date of the growing season, (2) the threshold temperature for accumulating the heat-sum units and (3) the critical heat sum total for the event (Boyer 1978). A heat sum equation must be fit to each latitudinal or physiographic region; no one equation including the generalized Boyer (1978) equation seems to fit the entire range of a single species.

In the case of *Pinus taeda* in North Carolina, heat sum predictions from Boyer (1978) adhere closely to actual pollen release, as shown in Fig. 5.5 (Williams 2008). Pollen shed dates for *Pinus taeda* begin in mid-February at the southerly latitudes in Texas, Louisana and Florida then extend upwards to the northerly latitudes of the species' range. Peak pollen shed from the northernmost part of the range, from North Carolina to Maryland usually occurs by mid-April (Baker and Langdon 1990).

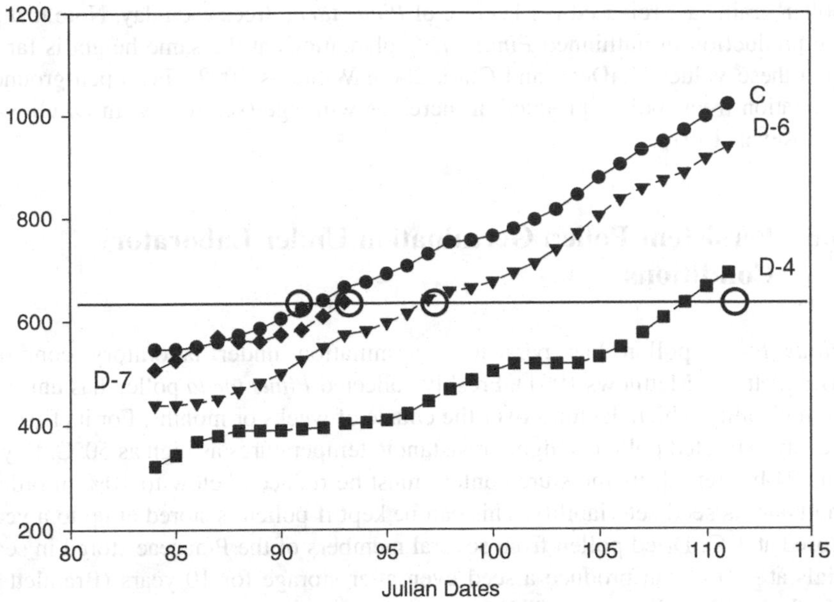

Fig. 5.5 The heat sum predictive equation (Boyer 1978) closely predicted peak pollen release for indigenous *Pinus taeda* seed sources planted in North Carolina's Piedmont and coastal regions. The Boyer (1978) equation had a baseline temperature of 55°F (or 12.78 C) and a heat sum accumulation start date of January 1. It assumes peak pollen shed would occur at the 636-h threshold for peak pollen shed (a straight line). This heat sum equation was tested at the Croatan National Forest at 34° 83′ W 76° 95′ N (coded as C) in 2006, peak pollen shed was predicted (and observed) at 91–92 Julian days. For the Blackwood Mountain plantation within the Duke Forest located at 35° 97′ W 79° 09′ W, peak pollen shed predictions also coincided with observed dates for 3 years: 2004 (coded as D-4), 2006 (coded D-6) and 2007 (not shown). Predicted dates of 111–112, 96–97, 91–92 (respectively) were a close fit to observed dates for peak pollen shed on 111–112, 96–97 and 93 Julian dates respectively. The horizontal line shows the cumulative heat sum threshold at which time the pollen is predicted to be released. Circles show actual pollen release. The closer the circles to the line, the closer the fit between predicted and observed. (From C. Williams. 2008. Aerobiology of *Pinus taeda* pollen clouds. Canadian Journal of Forest Research 38: 2177–2188. Copyright permission granted)

5.7 Quantity of Pollen Released

Heavy amounts of conifer pollen are released annually for a brief period of time in the spring. *Pinus taeda* pollen is released in vast quantities over a period of 10 days to 4 weeks (Blush 1986; Williams 2008). Peak pollen shed lasts even longer for *Pinus sylvestris* at boreal latitudes in northern Sweden (Lindgren et al. 1975).

As an example, an open-ground *Pinus taeda* tree at a height of 16 m will release 1,134 g of pollen over a 14-day pollen shed or roughly 81 g per tree per day in open-ground conditions (Parker and Blush 1996). On average, this translates into roughly 140,130 g of pollen per hectare each day. Using a microscope slide, the author has tallied roughly 1.3–1.5 × 10^6 *Pinus taeda* pollen grains per gram so, on average, 10^{11}

pollen grains are released per hectare of *Pinus taeda* trees each day. Note that pollen production in unthinned *Pinus taeda* plantations at the same height is far less than these values (LaDeau and Clark 2006; Williams 2008). For open-ground or plantation trees, pollen production increases with age (see review in Di-Giovanni and Kevan 1991).

5.8 Persistent Pollen Germination Under Laboratory Conditions

Pinus taeda pollen has persistent germination under laboratory conditions (Bramlett and Matthews 1991). Freshly collected *Pinus taeda* pollen has unusually high viability which declines over the course of weeks or months. For its first 24 h, freshly extracted pollen is highly resistant to temperatures as high as 50°C. Beyond this 24-h interval, its moisture content must be reduced below to 10% in order to maintain its seed-set viability. This can be kept if pollen is stored at up to a year if stored at 3°C. Dried pollen from several members of the Pinaceae stored in sealed vials at −20°C can produce a seed even after storage for 10 years (Bramlett and Matthews 1991; Jett et al. 1993).

5.9 Long-Distance Travel for Pine Pollen

Pine pollen also has an unusual degree of buoyancy relative to other seed plants. This has been demonstrated using a simple ballistics model where pine pollen is transported as far as 47–60 km within 3 h (Di-Giovanni and Kevan 1991; Di-Giovanni et al. 1996) and these long transport distances for conifer pollen have been reported for over 100 years. But the most extreme measurement point is still the pollen sampled in the Atlantic Ocean over 1,000 km from land by Gunnar Erdtman (Erdtman 1937).

Conifer pollen dispersal is bimodal (Hengeveld 1989; Nichols and Hewitt 1994; Nathan et al. 2002). Its first mode is local neighborhood dispersal (LND) which accounts for 99% of pollen or seeds (Clark et al. 1998) and this means that most pollen grains remain within the periphery of the source (Fig. 5.6).

The second mode is long-distance dispersal (LDD) which by definition accounts for a small fraction (e.g. 1%) of pollen or seeds (Hengeveld 1989; Clark et al. 1998). This mode occurs when seeds and pollen are vertically uplifted above the plant canopy by vertical eddies (Horn 2005) which act as turbulent ejection events (see Fig. 5.6). Once above the canopy, the seeds or pollen grains are caught by horizontal winds then move horizontally (Nathan et al. 2002). Described as such, the LDD dispersal process can account for the many anecdotal reports of long-distance conifer pollen movement from 3×10^1–10^3 km from source (Table 5.3).

Note too, that the wind speed increases with increasing distance from the earth's surface (Fig. 5.6). Horizontal wind speeds will be higher for a taller, mature forest

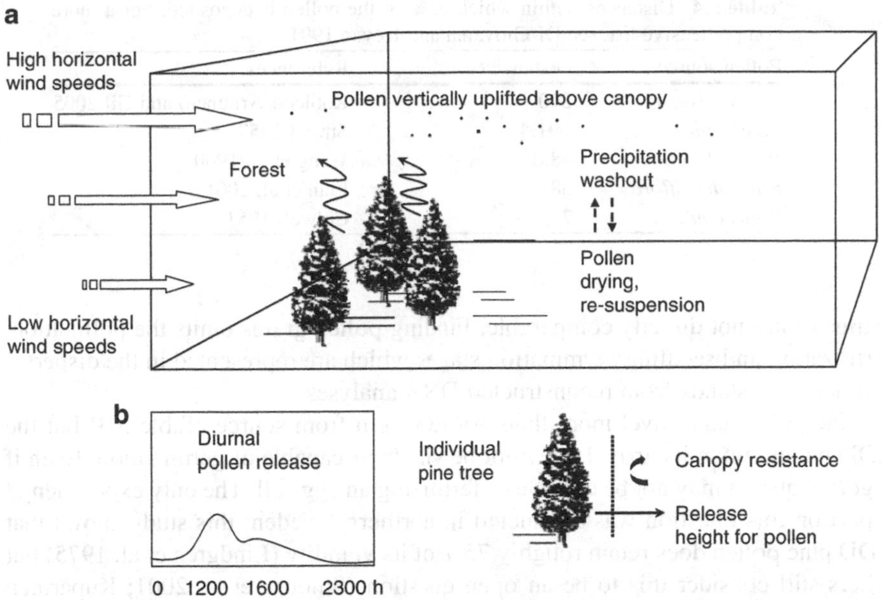

Fig. 5.6 The majority of released *Pinus taeda* pollen is deposited close to the adult sporophyte (local neighborhood dispersal or LND) but a tiny fraction is uplifted above the forest canopy by vertical eddies within the canopy then transported horizontally via long-distance dispersal (LDD) far from source (Figure modified from Williams 2008. Copyright permission granted)

Table 5.3 Maximum pollen dispersal distances reported for two genera within the Pinaceae family, *Pinus* spp. and *Picea* spp. (Modified from Williams 2008)

Pollen source	Location	Distance (km) from nearest source	Reference
Pinus spp.	Iowa USA	600	Bessey 1883
Pinus spp.	Gulf of Bothnia	30–55	Hesselman (1919) in Koski 1970
Picea spp.			
Pinus sylvestris	Southern Sweden	72	Review in Lanner 1966
Pinus spp.	Shetland Isles	250	Tyldesley 1973
Pinus spp.	Greenland	300	Rousseau et al. 2006
Picea spp.			

canopy than for a seedling canopy. Canopy attributes such leaf area index (LAI) provide canopy resistance and slow movement of released pollen grains. Airborne pollen within the canopy can be washed out of the atmosphere by rainfall or other heavy precipitation then dried and re-suspended (McDonald 1962). For *Pinus taeda*, typical pollen release peaks mid-morning between 1,000–1,200 h (Blush 1986). Another minor peak occurs in late afternoon but nighttime pollen concentrations for this species tend to be low or even absent (Blush 1986; Greenwood 1986; Williams 2008).

These anecdotal reports of remote pollen (Table 5.3) contrast sharply with dispersal distances reconstructed after the fact from DNA analyses (Table 5.4). These two

Table 5.4 Distances within which 93% of the pollen is deposited. For a more comprehensive list, see Di-Giovanni and Kevan 1991

Pollen source	Distance (m)	References
Pinus sylvestris	200	Robledo-Arnuncio and Gil 2005
Picea abies	>91.4	Strand 1957
Pinus elliottii	68.6	Wang et al. 1960
Pinus densiflora	68	Lian et al. 2001
Pinus coulteri	7	Colwell 1951

estimates are not directly comparable. Finding pollen grains omits the pollination, fertilization and seedling germination stages which are represented in the dispersal estimates for stands from reconstructed DNA analyses.

Pine pollen can travel more than 30–1,000 km from source (Table 5.3) but the LDD pollen, upon its arrival, may not be viable or capable of germination. Even if it germinates, it may not be capable of fertilizing an egg cell. The only experimental report on this question was conducted in northern Sweden; this study shows that LDD pine pollen does retain roughly 75% of its viability (Lindgren et al. 1975) but others still consider this to be an open question (Smouse et al. 2001; Kuparinen 2006). Laboratory estimates overestimate germination; field-captured pollen germination tends to be far lower as a rule.

The LND distances are collected from different types of studies so these shown in Table 5.4 are rough estimates at best. They tend to underestimate dispersal distances because they represent the distance for only 93% of the pollen. These distances are only on the scale of 5–200 m (Table 5.4).

Nonetheless, reporting LND and LDD movement separately in this fashion is consistent with the bimodal process of pollen dispersal although this practice runs contrary to the many DNA-based reports of average dispersal distance. Using averages as the dispersal measure is highly skewed in favor of LND events because LND events by definition are more frequent (99%). Another source of bias is that these also more likely to retain viability than LDD pollen (1%) and that DNA analyses sample only the recovered product of successful fertilization events. In principle, distances for LDD events can be recovered and estimated by sampling along the advection for historical wind movement patterns (Riechmann et al. 2006).

5.9.1 Measuring Terminal Velocity

How far the pollen particle moves depends on many factors including the particle's size, shape and velocity of sedimentation (settling or terminal velocity, V_t). Terminal velocity refers to the rate at which particles descend in still air owing to gravitational effects. When sedimentation is the only force responsible for deposition of particles, then the capture efficiency is simply the ratio between the vertical force (gravitational settling) and the horizontal force (horizontal wind speed).

Precise estimates hinge on directly measuring terminal velocity of the pollen particle itself. In still air, a single pollen grain falls slowly under gravity. The rate of its fall can be predicted or measured directly. Terminal velocity can be approximated for pollen and other spheroidal particles ranging in diameter from 1 and 70 μm using Stokes' Law:

Stokes' Law is described by:

$$V_t = \frac{2}{9}\frac{r^2 g[p-\sigma]}{\mu} \tag{5.1}$$

where V_t is terminal (or settling) velocity (cm s^{-1}), p is pollen density (g cm^{-3}), σ is the density of air (g cm^{-3}), r is the radius of the pollen grain (cm), g is the acceleration due to gravity (cm s^{-2}) and $\propto$ is dynamic viscosity of air (g cm^{-1} s^{-1}). Stokes' Law fits particles which are smooth spheroidal particles with diameters between 1 and 70 μm and this range approximates the size of conifer pollen (Jackson and Lyford 1999).

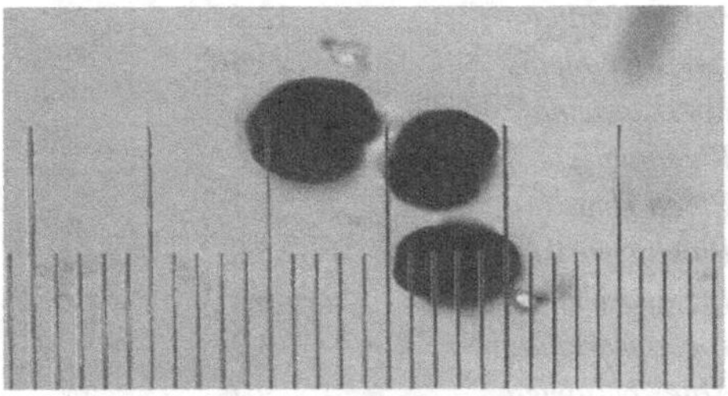

Photo 5.2 *Pinus taeda* pollen grains shown here are roughly 50 μm in breadth. Each gridline marks 10 μm. (Photograph from C. Williams. 2008. Aerobiology of *Pinus taeda* pollen clouds. Canadian Journal of Forest Research 38: 2177–2188. Copyright permission granted)

Pollen grain sizes are similar for *Pinus* spp. as shown by Cain (1940) who standardized measurements for these irregular shaped particles. Mean pollen grain breadth for *Pinus taeda* (Photo 5.2) is reported to be 53 μm (Cain 1940) which is similar to 58 μm for *Pinus strobus* pollen (Eisenhut 1961) yet *Pinus banksiana* pollen grains measure only 35 μm (Cain 1940). Members of the Cupressaceae have even smaller pollen grains measuring 10–35 μm (Jackson and Lyford 1999).

So how might one use these data to predict terminal velocity? As shown in Photo 5.2, a *Pinus taeda* pollen grain has a width of roughly 50 μm (Cain 1940; Eisenhut 1961; Di-Giovanni et al. 1995) so it falls within the range for Stokes' Law but its central capsule, with its two sacci, is not perfectly spheroidal. This shape deviation causes a downward bias for Stokes' Law predictions when compared to measured V_t values (Jackson and Lyford 1999).

This favors direct measurement over predictions (Table 5.5). Direct measurement of terminal velocity can be quite accurate because gymnosperms shed single pollen grains rather than compound or clumped pollen grains.

Terminal velocity is often measured using fall-towers (Box 5.2). Pollen is released from the top of a cylinder then timed before it reaches the bottom of the tower. Fall-towers vary in design and dimensions and this contributes to measurement error among studies (Eisenhut 1961; Di-Giovanni et al., 1995; Jackson and Lyford 1999; Aylor 2002). Despite this source of error, V_t values for many conifer

Table 5.5 Direct terminal velocity values for pollen among four genera within the Pinaceae: *Pinus, Larix, Picea, Abies*. Table modified from Williams, C.G.. 2008. Aerobiology of *Pinus taeda* pollen clouds. Canadian Journal of Forest Research 38: 2177–2188. Copyright permission granted

Species	Terminal velocity (V_t) (cm s^{-1})	Source
Pinus banksiana	3.1	Eisenhut 1961
Pinus banksiana	2.3	Di-Giovanni et al. 1995
Pinus taeda	2.3	Niklas 1984
Pinus taeda	2.1	Williams 2008
Pinus contorta	3.8	Eisenhut 1961
Pinus montana	3.3	Eisenhut 1961
Pinus nigra	4.5	Eisenhut 1961
Pinus parviflora	3.3	Eisenhut 1961
Pinus peuce	3.5	Eisenhut 1961
Pinus rigida	4.0	Eisenhut 1961
Pinus strobus	3.1	Eisenhut 1961
Pinus sylvestris	3.7	Eisenhut 1961
Larix decidua	12.6	Eisenhut 1961
Larix leptolepis	13.1	Eisenhut 1961
Larix laricina	3.1	Niklas 1984
Picea abies	5.6	Eisenhut 1961
Picea glauca	2.7	Niklas 1984
Picea omorika	5.2	Eisenhut 1961
Picea orientalis	6.1	Eisenhut 1961
Picea mariana	3.2	DiGiovanni et al. 1995
Abies balsamea	9.7	Eisenhut 1961
Zea mays	26.6	Aylor 2002
Zea mays'	31.0	Di-Giovanni et al. 1995

species range between 3 and 4 cm s^{-1} (Table 5.5) and the only notable exception is *Pinus taeda* which has a V_t value lower than other species, i.e. a value of 2.3 cm s^{-1} (Niklas 1984). This singular estimate was first obtained using a rectangular box but later falling tower measurements are similar (Table 5.5; Williams 2008). Using either protocol, *Pinus taeda* pollen has a terminal velocity value in the range of 2.1–2.3 cm s^{-1} which is low relative to other conifers.

Box 5.2 A fall-tower protocol for measuring terminal velocity of conifer pollen

This protocol, modified from the Aylor (2002) protocol, uses a clear glass tube in a darkened room. The experiment should be conducted in space which has high ceilings in addition to controlled temperature and humidity. The fall-tower is a 2.4-m glass settling tube which has an inner diameter of 0.02 m. The clear glass tube is illuminated using a cool fiber-optic light source at a distance of 0.3 m from the base of the glass tube. The top of the tube is covered with a thin diaphragm of aluminum foil which had a central pinhole of 0.2 mm diameter.

(continued)

Box 5.2 (continued)

The sample of pollen grains is placed here before being lightly tapped into the tube. The rate at which the pollen grains fall is measured at the start mark of 0.4 m from the top of the tube. The falling distance is 2 m.

Not all particles in a pollen sample are pollen grains. Fungal spores, other plant pollen and tapetal tissue debris are also present but too fine to eliminate before sampling. One way to cope is to identify the terminal velocity of the debris during a fall event; most debris particles are smaller than the large *Pinus* spp. pollen grains. To verify the stream of falling particles, a series of microscope slides, each with double-sided cellophane tape, are successively placed beneath the base of the tube during a single pollen fall event to determine which sample particles are pine pollen grains. Each microscope slide can be checked on site using a portable 10X light microscope prior to the terminal velocity measurements. Usually those slides which trapped slower particles show no pine pollen grains at all (Williams 2008).

Germination for the falling pollen particles can also be determined. Pollen is collected at base of the glass tube using Petri plates filled with 0.5% agar. The plates are then incubated for 48 h at 28°C before scoring germination tubes. Size, or diameter, of measured pollen grains can also be checked, as mentioned in Aylor's protocol. Measurements can be calibrated using a scale micrometer where each line marked 10 μm. Lastly, a portion of each pollen lot should be tested for pollen moisture content. The sample is weighted initially before drying it in a lab oven set at 50°C then re-weighed hourly until the sample weight no longer changed.

The most intuitive explanation for *Pinus* spp. pollen buoyancy is its two air-filled sacs or sacci attached to its otherwise round grain aid its airborne flight. The sacci are balloon-like, adding substantial surface area without adding much mass. And this is the case, as shown by recent computer simulation studies of pollen flight (Schwendemann et al. 2007). These authors show clearly that the sacci do act as airborne aids, decreasing terminal velocity of the pollen grain and thus increasing dispersal distance.

Another school of thought asserts that sacci aid only water flotation, not airborne flotation. The rationale is that sacci function as flotation devices in water (Doyle and O'Leary 1935; Tomlinson 1994; Runions and Owens 1999). To test this hypothesis, the latter authors compared two closely related species, one with sacci (*Picea abies*) the other without sacci (*Picea orientalis*). Only the saccate *Picea abies* pollen floats upwards in an aqueous solution for more than 1–3 min; *Picea orientalis* pollen sinks (Runions and Owens 1999). If the sacci aid water flotation but do not aid airborne flight then the two species should have similar terminal velocity values. And this is the case. *Picea abies* has a value of $Vt = 5.6$ and *Picea orientalis* has a value of $Vt = 6.1$ (Table 5.5).

But this comparison is uneven; *Picea abies* has a larger pollen grain size than *Picea orientalis*. A better comparison is the saccate *Picea omorika* which has pollen grains closer to size to the non-saccate *Picea orientalis*. From Table 5.3, we see that once the pollen grain sizes are similar, then terminal velocity values are no longer the same. *Picea omorika* saccate pollen will travel farther (*Vt* = 5.2) than the non-saccate *Picea orientalis* pollen (*Vt* = 6.1). This result implies that sacci add buoyancy for saccate pollen whether floating in water or in air.

But do sacci flatten upon desiccation? If so, this will alter the aerodynamics of pollen flight as well. Schwendemann et al. (2007) show that when pine pollen desiccates, its sacci close around the grain's distal aperture during flight (Schwendemann et al. 2007). This finding is incongruent for several reasons.

First, the pollen grain is already dessicated upon its release; a mature grain of the Pinaceae has less than 10% water content before being shed (Fernando et al. 2005). Second, others have observed that the central capsule of a *Pinus* spp. pollen grain keeps its shape and size during desiccation (Williams 2008). Its shape-retention property is consistent with the observation that the pine pollen grain's hard, thick exine retains its mouse-eared shape even after its central capsule is removed (Bohne et al. 2003).

Shape retention is not a trivial point when estimating aerodymanics of pollen flight. Consider the case of maize (*Zea mays*) pollen which shrivels into a non-spheroidal shape upon drying (Aylor 2002). Desiccated maize pollen has a *Vt* value of 31 but Aylor (2002) showed that fresh maize pollen has a value closer to 26 (Table 5.5). The terminal velocity measurements shown here for *Pinus taeda* assume that pine pollen is shape-constant whether dead or alive, fresh or desiccated (Williams 2008).

This assumption matters because small changes in terminal velocity can translate into large shifts in predicted dispersal distances. This can shown using ballistics models (Koski 1970; Di-Giovanni and Kevan 1991) and for more complex mechanistic models (Katul et al. 2006). For example, dispersal distance for a *Pinus* spp. pollen grain with a V_t value of 3.0 cm s^{-1} is predicted to move 49.6 km from source while a pollen grain with the higher value V_t value of 7.0 cm s^{-1} would travel only 26.8 km from source based on predictions simulated using turbulence conditions in a Durham NC pine plantation (Katul et al. 2006).

5.9.2 The Open Question of Long-Distance Pollen Germination

In practical terms, a pollen grain is only viable if it produces a seed, defined here as seed-set viability. Seed-set viability is not the same as germination although germination can be highly correlated with seed set (Bramlett and Matthews 1991). Harsh abiotic stresses such as extreme temperatures and high humidity coupled with ultraviolet (UV) radiation during pollen flight can reduce – or enhance – pollen germination. Effects for either freshly shed pollen or captured airborne pollen have not been reported to date.

By contrast, transient *Pinus spp*. pollen sampled in northern Sweden (Lindgren et al. 1975) and in Finland (Pulkkinen and Rantio-Lahtimaki 1995) had higher germination rates. These pollen germination data in Table 5.6 show almost no germination was observed.

Table 5.6 Low germination was recorded for *Pinus spp.* pollen collected on the Atlantic Ocean via the R/V Cape Hatteras ship from Beaufort North Carolina at the end of peak pollen shed in 2006. Treated Petri plates in Box 1 were exposed continuously from 0912 to 1700 h on 6 April. High germination rates for the control show that pollen placed into the closed agar-filled Petri plates inside each box received proper handling during the ship's voyage (From C. Williams, unpublished data)

Box	Plate	Pollen count	Pollen viability
1	1	10	0
1	2	85	0
1	3	18	0
1	4	76	0
1	5	108	1
Subtotal		297	0.30%
Control		84	96.4%

5.10 Variants in Microspore and Pollen Morphology for Conifers

In most conifers, male strobili are simple although male strobili in the non-coniferous gymnosperm *Gingko biloba* are compound. Spiral phyllotaxy for the microsporophylls is characteristic for the Pinaceae but the Cupressaceae have decussate phyllotaxy (Table 5.7). In all conifers, male strobili are grouped in panicles, racemes or clusters (Table 5.7).

Pollen development shows an unusual degree of variants: cell number, presence or absence of wings or sacci and if sacci are present, then the number of sacci vary

Table 5.7 A list of contrasting male characters between *Pinus* or Pinaceae and other conifers or gymnosperms (Character list adapted from Hart 1987)

Character	Taxon/taxa	Character	Contrast	Taxon/taxa
Male strobilus				
Arrangement	Pinaceae	Simple	Compound	*Gingko biloba*
Position	Pinaceae	Axillary	Terminal	Araucariaceae
Numbers	*Pinus*	Grouped in racemes or panicles	Grouped in clusters	*Pseudolarix*, *Keteleeria*
Phyllotaxy	Pinaceae	Spiral	Decussate	Cupressaceae
Microgametophyte				
Pollen sacs	*Pinus*, *Larix*, *Pseudotsuga*	Bisaccate	Three sacs per pollen grain	*Dacrydium*, *Microstrobus* (Podocarpaceae)
Prothallial cells	Pinaceae	1 or 2 cells	40	*Agathis* (Araucariaceae)
Sperm	Pinaceae	No walls	Cell walls	Cupressaceae
Sperm	Pinaceae	Equal	Unequal	Cupressaceae

(Table 5.7). Development of a microspore to a pollen grain can range from 4 weeks in *Pinus contorta* (Owens and Molder 1984) to as little as 7–11 days in *Pinus sylvestris* (Rowley et al. 2000). Tetrads form four microspores with two simultaneous cell divisions in *Pinus* spp. but other genera in the Pinaceae have sequential or bilateral cell divisions.

The number of mitoses during male gametophyte development differs widely among gymnosperm families; the Cupressaceae have only three cell divisions in the male gametophyte (microspore to tube vs. generative cells then two sperm cells complete with cell walls and organelles) but the Araucariaceae and Podocarpaceae have more than five. These and other families are compared in Fernando et al. (2005). Oddly, genera within the Podocarpaceae produce pollen with three sacci (Tomlinson 1994; Table 5.7) so the fluid dynamics of pollen dispersal for these taxa would provide an interesting study.

Some conifer taxa produce pollen with wings or sacci. Early in pollen grain development, the exine and intine separate to form the sacci. Some taxa have two sacci and others have three sacci. Once the microspore has formed its sacci, its central cell divides unequally twice, producing two small prothallial cells and a larger antheridial initial. Only the antheridial initial will divide, also unequally, to produce the small generative cell and the large tube cell (Owens and Molder 1984).

The number of cells in a single pollen grain at the time of its release varies among gymnosperm taxa. Two-celled male gametophytes are common in the Cupressaceae (Singh 1978; Chesnoy 1987). The two prothallial cells observed in the Pinaceae are absent. The first division of the microspore nucleus gives rise to only generative and tube cells in the Cupressaceae (Gifford and Foster 1989).

Table 5.7 shows *Pinus* spp. and a contrast for a few characters; a longer list of useful comparative characters, as given by Hart (1987) is only mentioned in part here: (1) male strobili are compound or simple, terminal or axillary, single at ends of leafy shoots, grouped in clusters or grouped in racemes and (2) male gametophyte has pollen tetrad formation which is simultaneous or successively bilateral; pollen grains are bisaccate, wingless or three sacs; prothallial cell numbers vary, sperm nuclei develop with or without cell walls and sperm cells are equal or unequal in size.

The male gametophyte of *Agathis* (Araucariaceae) continues cell division until 40 prothallial cells or nuclei are present. It is also the largest male gametophyte known in plants (Gifford and Foster 1989; Friedman 1993). By contrast, most angiosperm pollen has only two cells; the haploid microspore in angiosperms divides once to produce a vegetative cell and a generative cell. This is only one of many pollen biology characters which vary between conifers and other seed plants.

5.11 Closing

So how far does pine pollen move? Over 1,000 km from land, based on Gunnar Erdtman's pioneering experiments. These same experiments inspired our research group to test germination for pollen captured over the Atlantic Ocean. Germination was quite low in this preliminary experiment.

Microsporangia or pollen sacs attach to the abaxial surface of fertile scales on the strobilus. Sporogenesis takes place inside the pollen sacs. The end-product of meiosis is a tetrads of four microspores. Each microspore divides into a multicellular male gametophyte. All pollen has a siphonogamous pollen tube which emerges from a distal aperture. The microsporangiate strobilus and the pollen grain show a wide diversity of shapes, sizes and appendages among modern conifers. Less conserved characters for male reproduction in conifers include strobilus morphology, pollen sac number per scale, pollen size, number of prothallial cells, number of mitotic divisions and number of sacci (if any).

References

Aylor, D. 2002. Settling speed of corn (*Zea mays*) pollen. Journal of Aerosol Sciences 33: 1601–1607.

Baker, J. and O. Langdon. 1990. *Pinus taeda* L. Loblolly Pine. In: *Silvics of North America*, Volume 1, *Conifers*,. Agriculture Handbook 654. Forest Service. United States Department of Agriculture. Washington, DC. pp. 497–512.

Bessey, C. 1883. Remarkable fall of pine pollen. American Naturalist 17: 658.

Blush, T. 1986. Seasonal and diurnal patterns of pollen flight in a loblolly pine seed orchard. pp. 150–159. In: *Proceedings IUFRO Conference*, Williamsburg VA, October 13–17.

Bohne, G., E. Richter, et al. 2003. Diffusion barriers of tripartite sporopollenin microcapsules prepared from pine pollen. Annals of Botany 92: 289–297.

Bohne, G., H. Woehlecke, et al. 2005. Water relations of the pine exine. Annals of Botany 96: 201–208.

Boyer, W. 1978. Heat accumulation: An easy way to anticipate the flowering of southern pines. Journal of Forestry 76: 20–23.

Bramlett, D. and F. Bridgwater. 1989. Pollen development classification system for loblolly pine. In Proc 20th Southern Forest Tree Improvement Conference Charleston SC. pp. 116–121.

Bramlett, D. and F. Matthews. 1991. Storing loblolly pine pollen. Southern Journal of Applied Forestry 15: 153–157.

Cain, S. 1940. The identification of species in fossil pollen of *Pinus* by size-frequency determinations. American Journal of Botany 27: 301–308.

Chesnoy, L. 1987. La reproduction sexuée des Gymnospermes. Bulletin of the Botanical Society of France 134: 51–56.

Chichiricco, G. and E. Pacini. 2008. *Cupressus arizonica* pollen wall zonation and *in vitro* hydration. Pl. Syst. Evol. 270: 231–242.

Clark, J., S. Fastie, et al. 1998. Reid's paradox of rapid plant migration - Dispersal theory and interpretation of paleoecological records. Bioscience 48: 13–24.

Colwell, R. 1951. The use of radioactive isotopes in determining spore distribution patterns. American Journal of Botany 38: 511–523.

Dickinson, H. and P. Bell. 1976. Development of the tapetum in *Pinus banksiana* preceding sporogenesis. Annals of Botany 40: 103–113.

Di-Giovanni, F. and P. Kevan. 1991. Factors affecting pollen dynamics and its importance to pollen contamination: a review. Canadian Journal Forest Research 21: 1155–1170.

Di-Giovanni, F., P. Kevan, et al. 1996. Lower planetary boundary layer profiles of atmospheric conifer pollen above a seed orchard in northern Ontario, Canada. Forest Ecology and Management 83: 87–97.

Doyle, J. and M. O'Leary. 1935. Pollination in *Pinus*. Proceedings of the Royal Dublin Society 21: 181–190.

Engel, M., A. Chaboud, et al. 2003. Sperm cells of *Zea mays* may have a complex complement of mRNAs. Plant Journal 34: 697–707.

Eisenhut, G. 1961. Untersuchungen uber die Morphologie und Okologie der Pollenkorner heimischer und fremdlandischer Waldbaume. Forstwiss. Forsch. 15: 1–68.

Erdtman, G. 1937. Pollen grains recovered from the atmosphere over the Atlantic. Medd. Göteborgs Bot. Trädg 12: 185–196.

Fernando, D., J. Owens, et al. 2001. RNA and protein synthesis during in vitro pollen germination and tube elongation in *Pinus monticola* and other conifers. Sexual Plant Reproduction 13: 259–264.

Fernando, D., M. Lazarro, et al. 2005. Growth and development of conifer pollen tubes. Sexual Plant Reproduction 18: 149–162.

Frankis, R. 1990. RNA and protein synthesis in germinating pine pollen. Journal of Experimental Botany 41: 1469–1473.

Friedman, W. 1993. The evolutionary history of the seed plant male gametophyte. Trends in Ecology and Evolution 8:15–20.

Gifford, E. and A. Foster 1989. Morphology and evolution of vascular plants. New York, W.H. Freeman Company.

Greenwood, M. 1986. Gene exchange in loblolly pine: the relation between pollination mechanism, female receptivity and pollen availability. American Journal of Botany 73: 1443–1451.

Hart, J. 1987. A cladistic analysis of conifers: preliminary results. Journal of the Arnold Arboretum 68: 269–307.

Harrison, D. and M. Slee. 1992. Long shoot terminal bud development and the differentiation of pollen- and seed-cone buds in *Pinus caribaea* var. *hondurensis*. Canadian Journal of Forestry Research 22: 1565–1668.

Hengeveld, R. 1989. *Dynamics of Biological Invasions*. Chapman and Hall, London.

Hesselman, H. 1919. Iakttagelser över skogstradspollens spridningförmåga. Medd Skogsöfrsöksanst 16: 27–60.

Horn, H. 2005. Eddies at the gates. Nature 436: 179.

Jackson, S. and M. Lyford. 1999. Pollen dispersal models in Quaternary plant ecology: assumptions, parameters and prescriptions. Botanical Review 65: 39–75.

Jett, J., D. Bramlett, et al. 1993. Pollen collection, storage and testing. Editor: D.L. Bramlett. In: *Advances in Pollen Management, Agricutlural Handbook 698*. Government Printing Office, Washington DC. 101 p.

Katul, G., A. Poporato, et al. 2006. Mechanistic analytical models for long-distance seed dispersal by wind. American Naturalist 166: 368–381.

Katul, G., C. Williams, et al. 2006. Dispersal of transgenic conifer pollen. Editor: C.G. Williams. In: *Landscapes, Genomics and Transgenic Conifers*. Springer, Dordrecht, The Netherlands. pp. 121–143.

Koski, V. 1970. A study of pollen dispersal as a mechanism of gene flow in conifers. Comm. Inst. For. Fenn. 70: 1–78.

Kuparinen, A. 2006. Mechanistic models for wind-dispersal. Trends in Plant Sciences 6: 296–301.

LaDeau, S. and J. Clark. 2006. Annual pollen production in *Pinus taeda* grown under elevated CO_2. Functional Ecology 20: 541–547.

Lanner, R. 1966. Needed: a new approach to the study of pollen dispersion. Silvae Genetica 15: 50–52.

Lian, C., M. Miwa, et al. 2001. Outcrossing and paternity analysis of *Pinus densiflora* (Japanese red pine) by microsatellite polymorphism. Heredity 87: 88–98.

Lindgren, D., L. Paule, et al. 1975. Can viable pollen carry Scots pine genes over long distances? Grana 34: 64–69.

McDonald, J. 1962. Collection and washout of airborne pollen and spores of raindrops. Science 135: 435–437.

Nathan, R., G. Katul, et al. 2002. Mechanisms of long-distance dispersal of seeds by wind. Nature 418: 409–413.

Nichols, R. and G. Hewitt. 1994. The genetic consequences of long-distance dispersal during colonization. Heredity 72: 312–317.

Niklas, K. 1984. The motion of windborne pollen grains around conifer ovulate cones – implications on wind pollination. American Journal of Botany 71: 356–374.
Owens, J. and M. Molder. 1984. The reproductive cycle of lodgepole pine. Victoria, BC, Information Services Branch, British Columbia Ministry of Forests.
Parker, S. and T. Blush. 1996. Quantifying pollen production of loblolly pine (*Pinus taeda* L.) seed orchard clones. Westvaco Forest Research Report. 163 p.
Pettit, J. 1985. Pollen tube development and characteristics of the protein emission in conifers. Annals Botany 56: 379–397.
Pulkkinen, P., and A. Rantio-Lahtimaki. 1995. Viability and seasonal distribution patterns of Scots pine pollen in Finland. Tree Physiology 15: 515–518.
Reichman, J., L. Watrud, et al. 2006. Establishment of transgenic herbicide-resistant creeping bentgrass (Agrostis stolonifera L.) in nonagronomic habitats. Molecular Ecology 15: 4243–4255.
Robledo-Arnuncio, J.J. and L. Gil. 2005. Patterns of pollen dispersal in a small population of *Pinus sylvestris* L. revealed by total-exclusion paternity analysis. Heredity 94: 13–22.
Rousseau, D.-D., P. Schevin, et al. 2006. New evidence of long distance pollen transport to southern Greenland in late spring. Review of Palaeobotany & Palynology 141: 272–286.
Rowley, J., J. Skvarla, et al. 2000. Microsporogenesis in *Pinus sylvestris* L. VIII. Tapetal and late pollen grain development. Plant Syst. Evol. 225: 201–244.
Rudall, P. and R. Bateman. 2007. Developmental bases for key innovations in the seed-plant microgametophyte. Trends in Plant Science 12: 317–326.
Runions, C. and J. Owens. 1999. Pollination of *Picea orientalis* (Pinaceae): saccus morphology governs pollen buoyancy. American Journal of Botany 86: 190–197.
Schwendemann, A., G. Wang, et al. 2007. Aerodynamics of saccate pollen and its implications for wind pollination. American Journal of Botany 94: 1371–1381.
Singh, H. 1978. *Embryology of gymnosperms*. Berlin, Gebruder Borntraeger.
Smouse, P., R. Dyer, et al. 2001. Two-generation analysis of pollen flow across a landscape. I. Male gamete heterogeneity among females. Evolution 55: 260–271.
Strand, L. 1957. Pollen dispersal. Silvae Genetica 6: 129–136.
Tomlinson, P. 1994. Functional morphology of saccate pollen in conifers with special reference to the Podocarpaceae. International Journal of Plant Sciences 155: 699–715.
Tyldesley, J. 1973. Long-range transmission of tree pollen to Shetland. I. sampling and trajectories. New Phytologist 72: 175–181.
Wang, C.-W., T. Perry, et al. 1960. Pollen dispersion of slash pine (*Pinus elliottii* Englem.) with special reference to seed orchard management. Silvae Genetica 9: 78–86.
Williams, C. 2008. Aerobiology of *Pinus taeda* pollen clouds. Canadian Journal of Forest Research 38: 2177–2188.
Young, L. and R. Stanley 1963. Incorporation of tritated nucleosides thymidine, uridine and cystidine in nuclei of germinating pine pollen. Nucleus 6: 83–90.

Chapter 6
Pollination and Fertilization

Summary All conifers rely on wind to move pollen to ovule but form matters as much as chance; pollination is more akin to coordination and synchrony than it resembles a stochastic process. During female strobilus receptivity, ovules exude a localized pollination drop at night. By early morning, the pollination drop retracts, pulling its captured grains inside the micropylar arms, closer to the spongy nucellus. Hydrated by the pollination drop, the pollen grain now germinates into the spongy nucellar tissue. The pollen tube then halts its growth midway through the nucellus during the lengthy interval between pollination and fertilization. During this interval, the female gametophyte completes its development, slowly expanding to its maximum size and forming multiple archegonia. The duration of the pollination-fertilization interval is taxon-specific, lasting many months. The pollen grain resumes its growth a few days before fertilization then delivers one or two male gametes to the egg cell. The close synchrony between male and female reproduction is a sharp contrast to heterospory-induced divergence.

The nightly appearance – and disappearance – of the pollination drop is a mystery. This fluid extends beyond the micropylar arms of its ovule, picking up any deposited pollen grains and then retracts by early morning. The retracting drop deposits its hydrated pollen cache close to the spongy nucellus (Photo 6.1). What physical or chemical cues trigger the drop's withdrawal? Presumably its cue recognition system is localized because the drop is exuded by ovular tissues (O'Leary and von Aderkas 2006).

Contrary to popular opinion, the pollination drop is neither a water droplet nor a product of guttation in the adult tree (O'Leary and von Aderkas 2006). It is a localized phenomenon originating from the ovule's own sporophyte or gametophyte tissues. The drop is aqueous yet protein-rich. Its cues for cessation are thought to require particle size recognition, a chemical interaction or perhaps both. Pollen itself is thought to be the sole stimulus, not mechanical forces or evaporation (Tomlinson et al. 1997). These and other hypotheses have been further tested using *Juniperus communis*, a member of the Cupressaceae (Mugnaini et al. 2007).

Photo 6.1 *Pinus taeda* pollination drop from an ovule at the base of the cone scale (arrow) (Photograph taken by Floyd Bridgwater, USDA Forest Service. Permission granted)

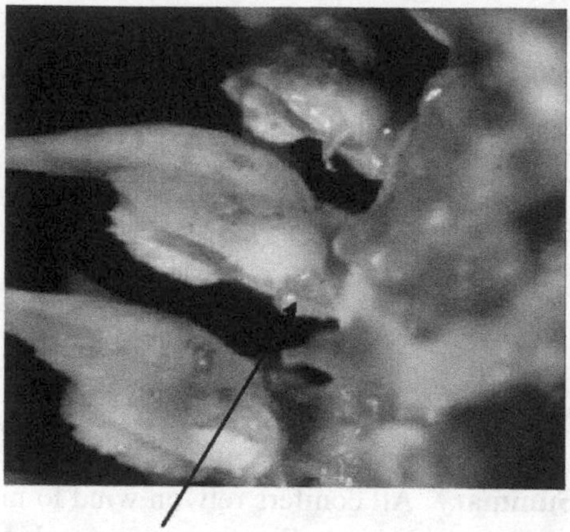

Juniperus communis provides an elegant *ex situ* system. The pollination period for this species is unusually long, lasting about as long as a month. A single female strobilus has three ovules and the diameter of each ovule's micropyle is 70 μm. Pollination drop emergence is not synchronous on each strobilus. Each drop can appear up to four times before an ovule's ability to form a drop is lost.

The experiments were conducted using branches with receptive female strobili collected on the previous evening. Short sprigs bearing female strobili were inserted into water-filled vials and kept under controlled conditions of 15°C with 52% relative humidity and the drop emerged. Drop volume was measured using a microcapillary tube. Particles were applied using a single human eyelash glued to a wooden stick with paraffin.

Particle applications to the pollination drop included desiccating silica particles in two sizes, small (10–15 μm) and large (63–200 μm), pollen from *Juniperus communis*, pollen from another conifer *Pinus canariensis*, pollen from an angiosperm *Pyrus communis* and live and heat-killed conspecific pollen (20 μm) (Mugnaini et al. 2007).

Experimental findings from this novel system offer new insights into the cuing mechanisms. First, particles are required for cuing; application of the eyelash itself, free of particles, did not alter pollination drop characteristics. Next, live *Juniperus communis* pollen triggered pollination drop withdrawal within 30 min, as expected. Third, the large silica particles raised the drop volume but did not cue its withdrawal. All other particles, including several types of pollen and small silica particles, triggered only a partial reduction in the drop volume. The authors reported that the partial drop withdrawal appears to be the drop's non-specific mechanical response to any small particle. Hydration was not ruled out as another explanation.

Total drop withdrawal, caused only by live conspecific pollen, is thought to be a two-part response to mechanical and molecular cues (Mugnaini et al. 2007).

To fully understand these experimental findings, it is necessary to take a more comprehensive view of pollination biology. Female strobilus receptivity is a logical starting place because this is where its coordination with the ovule and pollination drop begins.

6.1 Female Strobilus Receptivity

Female strobili become receptive to pollen entry when cone scales separate (Photos 6.1-6.2). The ovuliferous scales attached to the cone axis at an angle and the angle of the scales change with the stage of receptivity. If open, the angle favors pollen grains reaching the ovules.

As shown in Chapter 4, strobilus morphology has been divided into five stages (Pattison et al. 1969) but a six-stage classification system is more widely used (Bramlett and O'Gwynn 1980); both systems are based on degree of budbreak, strobilus elongation, size and distance between ovule-bearing scales.

Windborne pollen sifts between the open scales and some will land on the ovule's micropylar arms. Each pair of *Pinus taeda* ovules is located at the base of each fertile cone scale (Photo 6.1). Recall that the ovule has an inverted orienta-

Photo 6.2 A receptive *Pinus taeda* female strobilus from Bramlett and O'Gwynn (1980). Scales are starting to flex so that pollen can reach the pair of ovules located at base of each scale (Photographs by Floyd Bridgwater, USDA Forest Service. Permission granted)

tion so its micropylar arms hang down towards the cone axis. After the pollination drop emerges at night (Photo 6.1), it will retract, pulling pollen into the micropylar chamber (Photo 6.3A) where the hydrated pollen will germinate into the nucellar tissue.

After pollen capture, the drop will no longer emerge. The ovule closes its micropylar opening (Photo 6.3B). Ovuliferous scales of the female strobilus swell, sealing the entry to the ovules. This occurs even if the pollen grains do not germinate. This was the case for the ovule shown in Photo 6.3.

A *Pinus taeda* ovular opening is sealed at 4 weeks after pollination. The micropylar arms are sealed above the sharp outline of the nucellus even though the pollen grains did not germinate (Photograph by author)

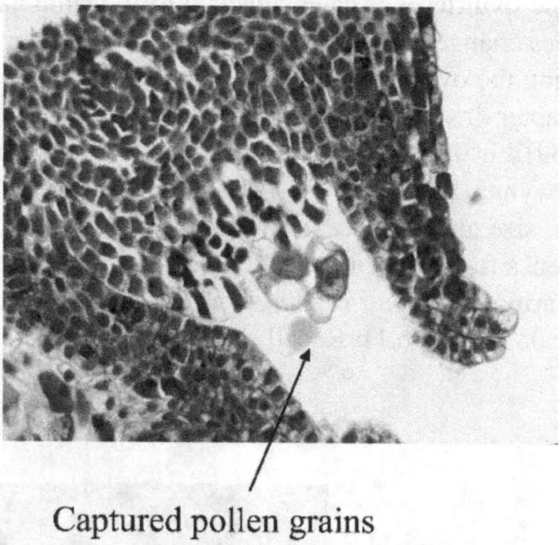

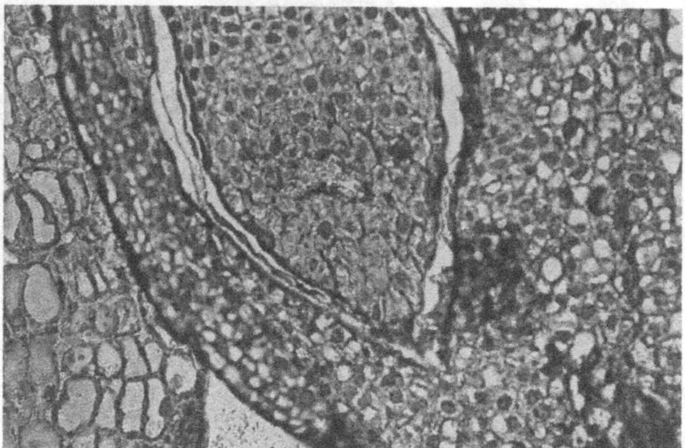

Photo 6.3A *Pinus taeda* pollen grains inside the micropylar chamber (arrow) just before germinating into the nucellar tissue of the ovule (Photograph taken by Floyd Bridgwater. Permission granted)

Photo 6.3B *Pinus taeda* ovule has sealed closed after pollination (Photograph taken by the author)

6.2 Pollination Drop: Localized Exudation from Each Ovule

The pollination drop is secreted by ovular tissues (McWilliam 1958). It could be a product of nucellar tissues although the integument and female gametophyte have also been suggested. Such definitive experiments on pollination drop origin have not yet been reported (see review by Gelbart and von Aderkas 2002). It is clear that the exudation is localized in ovular tissues. Recent experiments have clearly refuted the influences of guttation or xylem water potential changes in the adult sporophyte or high humidity in the atmosphere (O'Leary and von Aderkas 2006). The localized nature of the pollination drop is consistent with the lack of vascularized connections to the ovule (Singh 1978).

In *Pinus*, pollen capture is accomplished by means of a pollination drop exuded at night by the apex of the nucellus, filling the micropylar opening (Doyle and O'Leary 1935; Lee 1955; Tomlinson 1994). Secretion of the pollination drop starts at nightfall, reaching maximum exudation around 2 A.M. then recedes before daybreak. This emergence of the pollination drop is precise. Odlly, the drop is not secreted during daylight hours (Doyle and O'Leary 1935) nor in the presence of rain (Greenwood 1986; Brown and Bridgwater 1987).

When the pollination drop is reabsorbed, the pollen floats upwards into the ovule (Runions and Owens 1999) and transported to the surface of the nucellus (Brown and Bridgwater 1987). The pollination drop provides liquid for pollen hydration then deposits pollen at the nucellus to start tube growth through the maternal sporophyte tissue.

Few pollen grains reach the ovule although heavy quantities of pollen are released. Such an abundance of pollen leads to the presumption of allergies but conifer pollen rarely causes allergies. The unfortunate exceptions are a few members from the Cupressaceae family (Box 6.2).

6.3 Pollen Capture and the Role of the Micropyle

Most extant conifers have a pollination drop (Tomlinson et al. 1997; Owens et al. 1998; Gelbart and von Aderkas 2002). The notable exceptions include all *Abies* species and some *Tsuga* species. Some taxa rely on pollination drops exuded by the nucellus (Doyle and O'Leary 1935), others rely on pollination drops although rainwater also is an effective substitute (Greenwood 1986; Brown and Bridgwater 1987). Of the conifers, only members of the Araucariaceae completely lack a pollination drop (Gelbert and von Aderkas 2002).

The presence or absence of the pollination drop is only one character in a suite of correlated pollination characters among conifers (Tomlinson et al. 1997). Its absence correlates with germination of pollen outside the nucellus, defined as extended siphonogamy, found in *Tsuga* species and all members of the Araucariaceae. But it

Box 6.1 Zooidogamy and the pollination drop

The pollination drop was required for sperm delivery in the system of zooidogamy but now captures pollen (Labandeira et al. 2007). The drop once served the watery transport for motile antherozooids in the absence of a pollen tube (Fig. 6.1).

Only one living gymnosperm, *Gingko biloba*, has flagellated, motile sperm cells as well as the pollination drop – and a pollen tube (Fig. 6.1, stage b). Its pollen tube is branching and serves a haustorial function. The tube grows into the nucellar tissue like fungal hyphae then extracts nutrients for the gametophytic cells at the tube's growing end (Gifford and Foster 1989, p. 333).

The *Gingko biloba* pollen tube delivers two free-swimming sperm (Lee 1955) so the pollination drop is not required for sperm delivery. Each spermatozoid, including its ciliated tail, measures roughly 50–80 μm in length at release (Lee 1955). Upon their arrival, the egg cell first forms a small opening or beak at the top of its archegonium. A liquid forms then the first of the two sperm swim into the liquid. As soon as the sperm attaches to the egg, the beak of the egg retreats, making a path for only the head of the sperm so that most of the sperm body is left outside the archegonium. The egg forms a rigid membrane to prevent the entrance of the second sperm. The pollination drop, once so central to the prepollen delivery system, now provides the role of pollen scavenger.

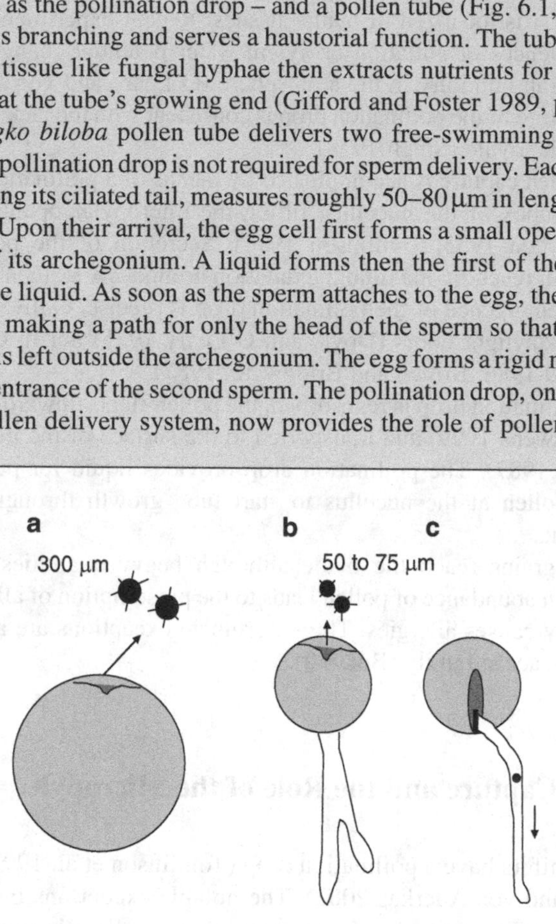

Fig. 6.1 The proposed evolutionary transition from prepollen to pollen transition, is redrawn from Poorts et al. (1996). Drawings represent prepollen and pollen without outer covering or exine: (**a**) male gametes by the late Paleozoic was large (300 μm) and zooidogamous, releasing motile spertherozoids through a proximal aperture, (**b**) from the late Paleozoic to early Mesozoic, a transitional form of zooidogamous prepollen is proposed which still releases motile antheroids through the proximal aperture but now has a haustorial pollen tube for nutritive functions only and (**c**) from Mesozoic to present-day, the siphonogamous pollen grain (50–75 μm) has a pollen tube which delivers immobile sperm nuclei (Pinaceae) or sperm cells (Cupressaceae) to the archegonium (From Poort et al. 1996. Copyright (1996) National Academy of Sciences, USA) Copyright permission granted

also correlates with pollen germinating inside the pollen chamber as in the case of *Larix* spp. and *Pseudotsuga* spp. (Tomlinson et al. 1997).

Its presence is the more common condition. This character correlates with (1) either saccate or non-saccate pollen and (2) bursting of pollen upon hydration (Tomlinson et al. 1997). The classification of correlated pollination characters was further expanded to five pollination types (Owens et al. 1998) based on a larger suite of traits including ovular morphology and orientation. It is interesting to note that one or more members of the Pinaceae are found in four of the five classification types, attesting to the considerable variation within a single family.

Another character, the delayed pollination drop, has now been included in this suite of characters (Gelbart and von Aderkas 2002). *Larix* spp. and *Pseudotsuga* spp, two genera in the Pinaceae, have delayed secretion where the drop appears weeks after pollination. In the case of *Pseudotsuga menziesii*, a secretion fills the micropylar canal about 7 weeks after pollination until fertilization (von Aderkas and Leary 1999). This phenomenon is better defined as a prefertilization drop rather than a pollination drop (Gelbart and von Aderkas 2002).

6.4 Composition and Function of the Pollination Drop

Pollination drops capture pollen but they might also provide pathogen protection. The pollen drop is rich in pathogenesis-related (PR) proteins including glucan-β-1,3-glucosidases (PR-2), chitinases (PR-3) and thaumatin-like proteins (PR-5) which degrade fungal cell walls and deter fungal growth (Wagner et al. 2007). Pollination drop may serve to retard fungal activity.

Box 6.2 Pollen allergens from the Cupressaceae family: a case of mistaken germination

Rarely does conifer pollen trigger allergies in humans but the notable exceptions are a few species within the Cupressaceae family. The well-documented species on the list includes *Juniperus ashei* in the southern central United States, *Cupressus sempervirens* in Italy, *Cupressus arizonica* in Spain and *Cryptomeria japonica* in Japan. Pollen from these species triggers an allergic response known as cedar fever. The pollen is inhaled then its epitopes enter the human blood stream and trigger an immune response. Certain air pollutants may heighten allergen expression; cedar fever affects nearly 10% of the exposed population in Japan yet nearly 20% of city dwellers are affected.

(continued)

Box 6.2 (continued)

The first question: what is the pollen biology behind these events? The pollen particles are small, ranging 15–35 µm in diameter, nonsaccate and star-shaped (Tomlinson 1994). The pollen grain enters the nose or mouth where it is hydrated by the moist mucosal lining, saliva or nasal fluids. The pollen grain starts to germinate into the mucosal lining. It bursts out of its thin exine within seconds then its thicker intine swells until the pollen grain becomes round rather than star-shaped. Now prepared to germinate, the pollen tube emerges from its aperture (Tomlinson 1994; Suarez-Cervera et al. 2003).

The second question: where is the allergen protein located? For *Cupressus sempervirens*, allergen proteins are highly concentrated on the pollen wall (including the intine) in addition to the central capsule's cytoplasm (Suarez-Cervera et al. 2003). When the pollen grain lands in the upper respiratory tract, its intine swells into a round shape bringing allergen proteins into contact with the human respiratory tract. Often, the pollen grain bursts then its rupture coats the allergen-rich cytoplasm contents on mucosal surfaces (Canini et al. 2004; Suarez-Cevera et al. 2003).

The third question: what is the identity and function of the candidate allergens? Oddly, the allergens are pathogen response (PR) proteins similar to those found in the pollination drop. Candidate *Jun a* 3 is a thaumatin-like protein (TLP) was isolated in *Juniperus ashei* (Midoro-Horiuti et al. 1999) and then used as a trans-specific probe to identify similar proteins in other conifers (Midoro-Horiuti et al. 2001; Suarez-Cervera et al. 2003; Cortegano et al. 2004).

As long suspected, air pollution enhances allergen expression (Cortegano et al. 2004) so humans living in urban settings will suffer cedar fever to a greater degree than those living in rural areas. It is not yet clear whether the carbohydrate moiety in *Jun a 3* glycoproteins also contributes to their immunoglobulin (Ig)E-binding capacity and ability to elicit IgE-mediated allergic symptoms (Breiteneder 2004). These and other similar questions are important to developing therapeutic relief from cedar fever.

Another candidate allergen isolated from *Juniperus ashei* is *Jun a 1*, a pectate lyase which degrades the pectin-rich intine, ensuring pollen tube germination (Suarez-Cervera et al. 2003). It has been characterized as a 40 kD glycoprotein (Midoro-Horiuti et al. 1999) with a full-length transcript of 1,101 nucleotides. Close relative *Juniperus virginiana* possesses an interesting mutation which reduces allergen response (Midoro-Horiuti et al. 2001).

Pollen from other conifers does not seem to induce these respiratory allergies but the reasons for this anomaly have not been explored. While it is true that *Pinus* spp. pollen does not burst nor shed its exine upon hydration, its PR-related allergens have not been localized nor characterized. This is also the case for other taxa which share this non-bursting or intact hydrated pollen character: *Larix*, *Pseudotsuga*, *Tsuga* as well as some of the Podocarpaceae (Tomlinson 1994).

(continued)

Box 6.2 (continued)

Pine pollen does not cause allergies because "it is too heavy to travel or that it falls beneath its source and therefore does not make its way up into the human nose." Long-distance dispersal of pine pollen has been well-established for over 100 years. Neither dispersal distance nor particle size determine whether conifer pollen triggers human allergies. A comparative analyses of candidate genes or gene products would be a better approach than relying on a medical myth.

6.5 Pollen Germination into the Nucellus

Once the pollen grain hydrates, its tube emerges. Immediate germination is the case for *Picea* and *Pinus* whether the pollen lands on agar, water, sugared water or nucellar tissue (Box 6.3). Other genera in the Pinaceae delay pollen germination by 3 weeks (*Pseudotsuga*), 3 months (*Keteleeria evelyniana*) or even as long as 9 months (*Cedrus* spp.) (Konar and Oberoi 1969).

Germination begins when the exine splits (Singh 1978, p. 132). In conifer taxa with sacci, the exine splits between the sacci at the distal end, along the suture or leptoma. Here, the hydrated pollen grain swells to the point that its sacci separate, exposing the leptoma. In *Pinus* spp. the pollen tube slowly emerges from the leptoma and begins its movement through the nucellus or spongy diploid ovular tissue. By contrast, germination starts on the cone (Tomlinson 1994). Other taxa germinate the pollen tube outside the ovule. Some pollen types in the Cupressacceae burst upon hydration to release the pollen tube (Box 6.2).

At this point, the tube nucleus migrates into the growing pollen tube and the generative cell divides equally to form two more cells: a fertile body cell and sterile stalk cell. The pollen tube begins its growth through the nucellus in various ways (Singh 1978, p. 137).

The simplest case is described for the Pinaceae. Intercellular signalling occurs between the growing pollen tube and the nucellus or other parts of the maternal sporophyte (Owens et al. 1990). This is consistent with experimental findings where angiosperm pollen tube growth depends on a calcium-mediated signal cascade as well as cues from haploid cells and diploid ovular tissue (Wilhelmi and Preuss 1999). In *Pseudotsuga menziesii*, signaling appears to start as early as 8 weeks after pollination (Takaso and Owens 1996).

The pollen tube elongates between nucellar cells. At this stage, the female gametophyte does not yet exist; only the megaspore mother cell (MMC) is present. The pollen tube exudes secretions which cause cell collapse including pectinase, cellulose, hydrolase, acid phosphatase, esterase, amylase, proteases and other degradative enzymes (Owens and Morris 1990). The pollen tube elongates through its own milieu of hydrolytic enzymes and the degenerating nucellar tissues.

Inside the elongating pollen tube of *Picea abies*, a well-organized network of microfilaments that extends the length of the tube (Terasaka and Niitsu 1994; Lazarro 1996, 1998). The microfilament network is orderly, forming two distinct zones within the elongating pollen grain. These zones partition plastids from mitochondria (Justus et al. 2004). Microfilaments in *Picea abies* pollen move in a fountain pattern, a pattern that is reversed in angiosperm pollen (Justus et al. 2004).

Box 6.3 Protocol for *Pinus taeda* pollen germination

This agar-based pollen viability assay (Goddard and Matthews 1981) is widely used for testing pollen prior to controlled pollinations.

1. Add 0.625 g agar to 125 ml distilled water to obtain a solution of 0.5% wt./vol. using Difco Bacto agar.
2. Sterilize agar in autoclave for 20 min.
3. Pour melted agar into petri dishes, filling sterile Petri dishes only ¼ full. Minimize exposure of agar plates at all stages to prevent microbial contamination.
4. After agar has solidified, dust re-hydrated pollen lightly on surface using a camel hair art brush. Use a different, clean art brush for each pollen sample.
5. Incubate dishes at 29°C for 48 h.
6. Using a dissecting microscope, tally at least 200 pollen grains per Petri dish. Only pollen grains with tubes equal to or exceeding the width of the grain are viable. Pollen germination above 80% germination is considered very good for stored pollen.

6.6 Pollen Tube Dormancy

The *Pinus taeda* pollen tube ceases growth partway through its germination into the nucellus. The pollen tube appears to be dormant. Dormancy of the pollen tube dormancy coincides with female meiosis, continues through the formation of the female gametophyte and does not break until a few days before fertilization. The pollen tube revives in response to some unknown cues. Candidates for cues include ovular secretions (Takaso and Owens 1996) or rapid female gametophyte growth (Gifford and Foster 1989). Other authors have suggested that this could be a stage suited to gametophytic selection (Takaso et al. 1995) but such evidence has yet to be recovered (Williams 2008). Similarly, archegonial development can be ruled out as the stimulus for resumption of pollen tube growth. In some conifers, pollen tube growth proceeds independently whether archegonia are dead or alive (Dumont-BeBoux et al. 1998).

6.7 Female Gametophyte Development After Pollination

All female reproductive cycles in conifers, regardless of duration, share a common feature: the female gametophyte develops *after* pollination in conifers. Its highly conserved development proceeds through three major stages: (1) a free nuclear phase, (2) a cellularization phase and (3) a cellular growth phase (Singh 1978; Konar and Moitra 1980; Friedman and Carmichael 1998). The breadth of variation among conifers and other gymnosperms has been reviewed in depth by Konar and Moitra (1980).

6.8 Fertilization

Once the pollen tube resumes elongation, it penetrates the megaspore wall in order to reach an archegonium (Pettit 1985). About a week before fertilization, its body cell divides to form two male nuclei which are unequal in size. The male gametes are nuclei or cells formed by the mitosis of the body cell.

In *Pinus*, the tip of a pollen tube forces itself between the neck cells of an archegonium and then ruptures, discharging the two male gametes, the tube nucleus and the sterile cell into the cytoplasm of the egg (Gifford and Foster 1989, p. 438; Owens and Morris 1990). The larger, leading sperm reaches the egg nucleus first and fertilization occurs (Runions and Owens 1999). The smaller sperm nucleus, the tube cell and the sterile cell now degenerate. In *Pseudotsuga menziesii*, microtubules associated with paternal organelles migrate with the leading sperm as it moves toward the egg nucleus (Owens and Morris 1990).

The members of the Cupressaceae have equally-sized male cells, rather than nuclei, which form much later when the pollen tube enters the archegonial chamber (Singh 1978). The number of sperm cells varies among these taxa although two sperm cells (defined as diplospermy) is common. An exceptional case has been reported for *Cupressus arizonica* which is reported to have 12–14 sperm cells (Doak 1932).

Conifers do have a few rare cases of multiple fertilization. A single *Callistris* pollen tube can deliver multiple sperm to more than one archegonium (Baird 1953; Willson and Burley 1983). On rare occasions, two separate sperm fertilize an egg nucleus and a second cell nucleus within the same archegonium (Friedman 1992).

6.9 Different Female Reproductive Cycles

Conifer reproduction is synchronous with seasonal change in temperate zones. Reproductive development slows to a halt during winter then resumes each spring. The cycle, punctuated by seasonal change, can take 1, 2 and even 3 years from pollination to seed maturation.

Conifers are classified as 1-, 2- or 3-year reproductive cycles (Singh 1978). The cycle refers to the completion of female strobilus development from initiation to seed maturation. By comparison, male strobilus development is completed in a single year (Singh 1978) regardless of the duration of its respective female reproductive cycle.

All three types of reproductive cycles have a lengthy gap between pollination and fertilization, another feature that distinguishes gymnosperms from angiosperms (Fernando et al. 2001). As described in the following section, 3-year reproductive cycle is a heterogeneous grouping.

6.9.1 One-Year Reproductive Cycle

The genera in this group include *Abies*, *Picea*, *Cedrus*, *Pseudotsuga*, *Tsuga*, *Keteleeria* (Pinaceae) and *Cupressus*, *Thuja*, *Cryptomeria*, *Cunninghamia* and *Sequoia* (Cupressaceae). As an example, female strobili are initiated in late summer or fall of 2000 then they overwinter. Female strobili emerge followed by pollination in spring 2001. Fertilization takes place in summer of 2001, only 3–4 months after pollination (Singh 1978, pp. 245–246). Cones mature and seeds are then shed by the end of 2001. Pollination and fertilization occurs within the same year in a single growing season. The pollination-fertilization interval for the 1-year cycle is measured in months, not years.

6.9.2 Two-Year Reproductive Cycle

The genera included here are *Widdringtonia*, *Sequoiadendron* (Cupressaceae) and most species of *Pinus*. Female strobilus initials are formed in late summer or fall of 2000 then overwinter. In spring 2001, female strobili emerge, receive pollen in the first spring 2001 and become conelets. The conelet goes through another winter rest and in spring 2002, archegonia form in the conelet. Fertilization of the archegonia occurs by early summer of 2002 so the pollination-fertilization interval exceeds a year. After fertilization, the conelet is considered an immature cone. Maturation occurs by autumn 2002 at which time seeds are shed. Note that in this case, the 1-year and the 2-year cycles differ mainly in the duration of the pollination-fertilization interval (Singh 1978, pp. 246–247).

6.9.3 Three-Year Reproductive Cycle

Very few conifer species have a 3-year cycle but even so, all of these species do not share the same pollination-fertilization interval. Three of these are pine species (*Pinus pinea*, *Pinus leiophylla*, *Pinus torreyana*) which have pollination and fertilization events separated by a 2-year interval. As an example, female strobili initiated during late summer or autumn 2000 overwinter until spring 2001. Female strobili emerge then pollination occurs in spring 2001 then the pollinated strobili become conelets in 2001. The female gametophytes in the conelet develop so slowly that the megaspore does not go through free-nuclear divisions until autumn 2002. The conelet then overwinters again in the free-nuclear female gametophyte stage. Fertilization takes place by early summer 2003 and seeds mature in the cones by autumn 2003 (Singh 1978, p. 249). By contrast, the pollination-fertilization the interval varies for other species in this group. For example, *Juniperus communis* takes only one year (Ottley 1909) but for *Callitris robusta*, the pollination-fertilization interval takes 18 months (Baird 1953).

6.10 Closing

The mystery of the nocturnal pollination drop has yet to be completely solved but recent experiments have contributed substantially. Pollination can be seen as the convergence of opposing selective forces: (1) heterospory-induced divergence for male and female reproductive morphology versus (2) a precisely coordinated synchrony between female and male reproductive development as both move towards the singular goal of fertilization. The best example of this convergence is the pollination drop itself. It emerges at this critical female-male juncture, the coinciding of female strobilus receptivity and pollen shed. The drop captures pollen, hydrates pollen then positions pollen next to the spongy nucellus. Finding the pollination drop's cues opens an interesting research topic.

Conifers show a range of interesting variants on the wind-pollination systems. In some taxa, pollen capture is followed by immediate germination while others have germination delays that can last weeks or months. Most pollen germinates inside the ovule but some taxa have pollen that germinates outside the ovule. Other taxa have pollen which floats, sinks or bursts prior to pollen tube emergence. Multiple archegonia located at the micropylar end is the common condition but other taxa form large archegonial complexes and a few form archegonia at the chalazal end. Male gametes can be sperm nuclei, sperm cells or free-swimming sperm.

The female reproductive cycle spans either 1, 2 or 3 years in duration. Its duration is mostly defined by one component, its pollination-fertilization interval. In all cases, female gametophytes slowly develop well after pollination. In all cases, the elongating pollen tube is dormant during the pollination-fertilization interval. Male and female gametophyte development are independent, proceed on different timetables yet converge at the highly integrated, complex event of fertilization.

The role of the sporophyte dominates all aspects of the wind-pollination system. From strobilus initiation to pollination, the adult sporophyte develops the ovular tissues, opens the female strobilus, captures pollen grains on the micropylar arms then provides the nucellar medium for pollen germination. The female gametophyte has not yet developed up until this point hence it exerts no known functional role (other than perhaps the pollination drop). Just prior to fertilization, the roles of sporophyte and its endosporic female gametophyte suddenly switch. The female gametophyte now becomes dominant, differentiating multiple archegonia and triggering renewed pollen tube growth. Pollination particulars show the intricacy of the monosporangiate system.

References

Baird, A. M. 1953. A life history of *Callitris*. Phytomorphology 3: 258–284.

Bramlett, D. and C. O'Gwynn. 1980. Recognizing developmental stages in southern pine flowers: the key to controlled pollinations, USDA Forest Service Southeastern Experiment Station, 14 pages.

Brown, S. and F. Bridgwater. 1987. Observations on pollination in loblolly pine. Canadian Journal of Forest Research 17: 299–303.

Breiteneder, H. 2004. Thaumatin-like proteins – a new family of pollen and fruit allergens. Allergy 59: 479–481.

Canini, A., J. Giovinazzi, et al. 2004. Localisation of a carbohydrate epitope recognised by human IgE in pollen of Cupressaceae. Journal of Plant Research 117: 147–153.

Cortegano, I., E. Civantos, et al. 2004. Cloning and expression of a major allergen from *Cupressus arizonica* pollen, *Cup a 3*, a PR-5 protein expressed under polluted environment. Allergy 59: 485–490.

Dumont-BeBoux, N., W. Weber, et al. 1998. Intergeneric pollen-megagametophyte relationships of conifers in vitro. Theor. Appl. Genet. 97: 881–887.

Doak, C. 1932. Multiple male cells in *Cupressus arizonica*. Botanical Gazette 94: 168–182.

Doyle, J. and M. O'Leary. 1935. Pollination in *Pinus*. Sci. Proc. Roy. Dublin Soc. 21: 181–190.

Fernando, D., J. Owens, et al. 2001. RNA and protein synthesis during in vitro pollen germination and tube elongation in *Pinus monticola* and other conifers. Sexual Plant Reproduction 13: 259–264.

Friedman, W. 1992. Double fertilization in nonflowering plants and its relevance to the origin of flowering plants. International Review of Cytology 140: 319–355.

Gelbart, G. and P. von Aderkas. 2002. Ovular secretions as part of pollination mechanisms in conifers. Annals of Forest Science 59: 345–357.

Gifford, E. and A. Foster. 1989. *Morphology and Evolution of Vascular Plants*. W.H. Freeman, New York.

Goddard, R. and F. Matthews. 1981. Pollen testing. Editor: E.C. Franklin. In: *Pollen Management Handbook*. Agricultural Handbook 587, Washington, DC, pp. 40–43.

Greenwood, M. 1986. Gene exchange in loblolly pine: the relation between pollination mechanisms, female receptivity and pollen viability. American Journal of Botany 73: 1433–1451.

Justus, C., P. Anderhag, et al. 2004. Microtubules and microfilaments coordinate to direct a fountain streaming pattern in elongating conifer pollen tube tips. Planta 219: 103–109.

Konar, R. and A. Moitra. 1980. Ultrastructure, cyto- and histochemistry of female gametophyte of gymnosperms. Gamete Research 3: 67–97.

Konar, R. and Y. Oberoi. 1969. Recent work on reproductive structures of living conifers and taxads - a review. Botanical Review 35: 89–116.

Labandeira, C., J. Kvacek et al. 2007. Pollination drops, pollen and insect pollination of Mesozoic gymnosperms. Taxon: 56: 663–695.

Lazarro, M. 1996. The actin microfilament network within the elongating pollen tubes of the gymnosperm *Picea abies* (Norway spruce). Protoplasma 194: 186–194.

Lazarro, M. 1998. The spermatagenous body cell of the conifer body cell of the conifer *Picea abies* (Norway spruce) contains actin microfilaments. Protoplasma 201: 194–201.

Lee, C. 1955. Fertilization in *Gingko biloba*. Botanical Gazette 117: 79–100.

McWilliam, J. 1958. The role of micropyle in the pollination of *Pinus*. Botanical Gazette 120: 109–117.

Midoro-Horiuti, T., R. Goldblum, et al. 1999. Molecular cloning of the mountain cedar (*Juniperus ashei*) pollen major allergen, *Jun a 1*. Journal of Allergy and Clinical Immunology 104: 613–617.

Midoro-Horiuti, T., R. Goldblum, et al. 2001. Identification of mutations in the genes for the pollen allergens of eastern red cedar (*Juniperus virginiana*). Clinical and Experimental Allergy 31: 771–778.

Mugnaini, S., M. Nepi, et al. 2007. Pollination drop in *Juniperus communis*: response to deposited material. Annals of Botany 100: 1475–1481.

O'Leary, S. and P. von Aderkas. 2006. Postpollination drop production in hybrid larch is not related to the diurnal pattern of xylem water potential. Tree 20: 61–66.

Ottley, A. 1909. The development of the gametophytes and fertilization in *Juniperus communis* and *Juniperus virginiana*. Botanical Gazette 48: 31–46.

Owens, J. and S. Morris. 1990. Cytological basis for cytoplasmic inheritance in *Pseudotsuga menziesii*. I. Pollen tube and archegonial development. American Journal of Botany 77: 433–445.

Owens, J., T. Takaso, et al. 1998. Pollination mechanisms in conifers. Trends in Plant Science 3: 479–485.

Pattison, J., J. Burley, et al. 1969. Development of the ovule strobilus in *Pinus kesiya* Royle ex Gordon (syn. *P. khasya Royle*) in relation to controlled pollination in Zambia. Silvae Genetica 18: 108–111.

Pettit, J. 1985. Pollen tube development and characteristics of the protein emission in conifers. Annals of Botany 56: 379–397.

Poort, R., H. Visscher, et al. 1996. Zoidogamy in fossil gymnosperms: the centenary of a concept, with special reference to prepollen of late Paleozoic conifers. Proceedings National Academy of Sciences USA 93: 11713–11717.

Runions, C. and J. Owens. 1999. Sexual reproduction of interior spruce (Pinaceae). I. Pollen germination to archegonial maturation. Intl. J. Plant Sci. 160: 631–640.

Singh, H. 1978. *Embryology of gymnosperms*. Berlin, Gebruder Borntraeger.

Suarez-Cervera, M., Y. Takahashi, et al. 2003. Immunocytochemical localization of *Cry j 1*, the major allergen of *Cryptomeria japonica* (Taxodiaceae) in *Cupressus arizonica* and *Cupressus sempervirens* (Cupressaceae) pollen grains. Sexual Plant Reproduction 16: 9–15.

Takaso, T. and J. Owens. 1996. Effects of ovular secretions on pollen in *Pseudotsuga menziesii* (*Pinaceae*). American Journal of Botany 81: 504–513.

Takaso, T., P. von Aderkas, et al. 1995. Prefertilization events in ovules of *Pseudotsuga*: ovular secretion and its influence on pollen tubes. Canadian Journal of Botany 74: 1214–1219.

Terasaka, O. and T. Niitsu. 1994. Differential roles of microtubules and actin-myosin cytoskeleton in the growth of *Pinus* pollen tubes. Sexual Plant Reproduction 7: 264–272.

Tomlinson, P. 1994. Functional morphology of saccate pollen in conifers with special reference to the Podocarpaceae. International Journal of Plant Sciences 155: 699–715.

Tomlinson, P., J. Braggins, et al. 1997. Contrasted pollen capture mechanisms in Phyllocladaceae and certain Podocarpaceae (Coniferales). American Journal of Botany 84: 214–223.

von Aderkas, P. and C. Leary. 1999. Micropylar exudates in Douglas fir – timing and volume of production. Sexual Plant Reproduction 11: 354–356.

Wagner, R., S. Mugnaini et al. 2007. Proteomic evaluation of gymnosperm pollination drop proteins indicates highly conserved and complex biological functions. Sexual Plant Reproduction 20: 181–189.

Williams C. 2008. Selfed embryo death in *Pinus taeda*: a phenotypic profile. New Phytologist 178: 210–222.

Willson, M. and N. Burley. 1983. *Mate choice in plants*. Princeton NJ, Princeton University Press.

Wilhelmi, L. and D. Preuss. 1999. The mating game: pollination and fertilization in flowering plants. Current Opinions in Plant Biology 2: 18–22.

Chapter 7
Syngamy, Embryo Development and Seed Dispersal

Summary The basic plan in modern conifers is conserved through organelle exclusion and zygote formation, proembryo development, early embryogeny and late embryogeny. Virtually all conifers have some type of polyembryony but not all have multiple archegonia. The ovule, once fertilized, develops into a seed. Each developing embryo is nourished by the female gametophyte as it grows out of its archegonium into the corrosion cavity. If it survives intense competition to become the dominant embryo, then it will develop a root meristem elongating in the direction of the micropyle and a shoot meristem elongating in the direction of the chalaza. A single, dominant embryo reaches maturity prior to seed dispersal. Seed dispersal occurs by wind but humans, birds and other vertebrates play an important role too.

Edible pine seeds are highly prized as a tasty food source but they also provide a quick insight into conifer seed development. Open one of these seeds. Inside is a single developing embryo nested inside its female gametophyte. The sculpted shape of the embryo – and its hollow space inside its female gametophyte – is not only the product of cellular replication but also the product of programmed cell death (PCD). Programmed cell death refers to the genetically-controlled, orderly process ensuring complete cellular degradation. It is a pathway of expressed genes leading to cellular suicide (Lam 2004). Dying cells exhibit a suite of canonical features including internucleosomal cleavage of nuclear DNA into fragments, cytoplasm shrinkage and chromatin condensation (Hiratsuka et al. 2002; Lam 2004). That edible pine seed once had multiple embryos but all embryos except for the dominant embryo succumbed to an early death. Likewise, the hollow V-shaped corrosion cavity in the female gametophyte was formed by programmed cell death earlier in embryo development.

Which tissue initiates the signal to die? This is a complex question because a developing seed is composed of parts from the entire diplohaplontic life cycle. The molecular recognition or signalling between gametophyte and sporophyte phases is essential to understanding this complex life cycle yet it is still an unexplored research frontier.

C.G. Williams, *Conifer Reproductive Biology*,

The following chapter covers the pre-fertilization interaction between male and female gametophytes: delivery of sperm to the egg cells, syngamy, organelle sorting, mitosis and early zygote formation. From there, polyembryony patterns and embryo development of the young sporophyte are presented. Seed dispersal and a brief history of the edible pine seed complete the chapter's contents.

7.1 Syngamy and Organelle Sorting

The release of two sperm (diplospermy) at the time of fertilization is a conserved character for conifers, gymnosperms and other seed plants. In *Pinus* species, both sperm enter the egg cell but only one, defined as the leading sperm, fuses with the egg nucleus. As the leading sperm nucleus approaches the egg nucleus, the egg cytoplasm is dense with maternal mitochondria and ribosomes (Owens and Morris 1990). The sperm nucleus settles into the egg nucleus, forming a cup-shaped depression. A double membrane separates the male and female nuclei (McWilliam and Mergen 1958) although their respective nuclear envelopes do fuse at several points. The paternal organelles remain separated from maternal organelles during this stage (Owens and Morris 1990).

Maternal and paternal sets of chromosomes now align on an equatorial plate (McWilliam and Mergen 1958). Nuclear membranes dissolve and disappear at the first mitotic division. The fused nucleus appears large and dense. The first mitosis of the zygote nucleus occurs within 24 h of fertilization, splitting the cell into two free nuclei (Runions and Owens 1999).

The first cell division after syngamy is the critical step for the mechanism of organelle sorting. The subsequent events depend on the pattern of organelle inheritance; conifers have at least two or perhaps three different patterns. Two of these patterns of organellar inheritance in gymnosperms are described in Table 7.1. The third pattern (not shown) is specific to *Gingko biloba* which is hypothesized to have strictly maternal inheritance of organelles (Mogensen 1996).

In general terms, the two free nuclei, one from each parent, are suspended in cytoplasm while organelle sorting, exclusion and transmission events are completed. These events are complete by the second mitotic division during which a total of four free nuclei are produced. These nuclei descend to the base of the egg

Table 7.1 Organelle inheritance patterns for gymnosperm families. Most angiosperms have maternal organelle inheritance (Mogensen 1996). Paternal (P) and maternal (M) organelle inheritance patterns differ among conifers

Taxa	Plastids	Mitochondria	References
Pinus taeda (Pinaceae)	P	M	Neale and Sederoff 1989
Taxus baccata (Taxaceae)	P	M	Pennell and Bell 1988
Agathis robusta (Araucariaceae)	P	P	Kaur and Bhatnagar 1984
Calocedrus decurrens (Cupressaceae)	P	P	Neale et al. 1991
Cryptomeria japonica (Cupressaceae)	P	P	Ohba et al. 1971

and initiate zygote formation (McWilliam and Mergen 1958). Subsequent mitoses leading to zygote formation now proceed. Each new cell has its correct complement of its maternal or paternal organelles.

7.1.1 Inheritance of Maternal Mitochrondria (M) and Paternal Plastids (P) for the Pinaceae and the Taxaceae

Biparental organelle inheritance for the Pinaceae and the Taxaceae families proceeds as follows. Here, mitochondria are inherited from the maternal parent and plastids are inherited from the paternal parent (Table 7.1). The mechanism has been described as three events:

1. Maternal mitochondria are retained in the distinct perinuclear zone.
2. Paternal mitochrondria and paternal plastids move with sperm nuclei into the egg cytoplasm (Owens and Morris 1990). Other authors describe the transfer of paternal plastids from the cytoplasm of the pollen grain's body cell to the egg cytoplasm (Guo et al. 1999).
3. Finally, maternal plastids become large inclusions which are excluded from the zygote.

Only at the first cell division do the paternal plastids and maternal mitochrondria surround each of two resulting nuclei (Runions and Owens 1999). Both paternal plastids and paternal mitochrondria are still present at syngamy in *Picea* spp. but not after the first cell division (Runions and Owens 1999).

Organelle exclusion is tightly orchestrated yet leaky (Owens and Morris 1990). This has been confirmed by DNA-based paternity analysis using mitochrondrial DNA from *Pinus banksiana* and *Pinus contorta*. Roughly 6% of the viable seedlings showed aberrant organelle inheritance (Wagner et al. 1991).

7.1.2 Inheritance of Paternal Mitochrondria (M) and Paternal Plastids (P) for the Cupressaceae and the Araucariaceae

Strictly paternal organelle inheritance occurs in some taxa within the Cupressaceae and Araucariaceae families (Table 7.1). Here, organelle exclusion proceeds through a different trio of events:

1. The maternal organelles in the egg cell are distributed throughout the cytoplasm which lacks a perinuclear zone.
2. The sperm nucleus, now enclosed in a cell wall, is surrounded by its own dense organelle-rich cytoplasm. At the union of sperm and egg cells, the male

cytoplasm surrounds the zygote nucleus so that only paternal plastids and paternal mitochrondria are proximal to the zygote.

3. Maternal mitochondria and maternal plastids degenerate at this point (Mogensen 1996).

Two or three different organelle inheritance patterns among conifers and other gymnosperms attests to the surprising diversity found among a few Mesozoic lineages.

7.2 Two Types of Polyembryony

Another peculiar feature of conifer reproductive biology is polyembryony or multiple embryos within a single seed. Contrary to popular usage, multiple archegonia and polyembryony are not interchangeable terms. This is because rarely are all archegonia fertilized and even so, there is more than one type of polyembryony. Multiple archegonia lead to polyzygotic embryos but cleavage polyembryony also occurs when a fertilized egg splits. This distinction is addressed here in more detail:

Simple polyembryony. Each embryo comes from a separate fertilization event. Each archegonium in an ovule is generally fertilized by a different pollen grain, although some interesting exceptions have been noted for the Cupressaceae (Doak 1932). Polyzygotic or archegonial embryos are the consequence.

Another source of confusion here is the degree of relatedness between polyzygotic embryos within a single seed (Williams 2007). They are more related than full-sibs but less related than identical twins.

Cleavage polyembryony. The second type of polyembryony, cleavage polyembryony, occurs when a single zygote splits into multiple embryos. These embryos are genetically identical. Knowing whether the species in question has one or two types of polyembryony drives the selective consequences of polyembryony, if any exist.

7.3 Four Patterns of Polyembryony

Again, conifers have a wide assortment of polyembryony patterns even within a single taxonomic family. As shown in Table 7.2, patterns of polyembryony vary among genera or even species within a family (Roy Chowdhury 1962; Dogra 1967; Singh 1978, pp. 208–209).

7.3.1 Pattern of Both Simple (S) and Cleavage (C) Polyembrony

Cedrus, *Pinus*, *Keteleeria* and *Tsuga* (Pinaceae) all show simple and cleavage polyembryony (Konar and Oberoi 1969). In *Pinus radiata*, cleavage embryos are

Table 7.2 Four polyembryony patterns for conifers and other gymnosperms

Polyembryony patterns	Examples
Simple (S) and Cleavage (C) polyembrony	*Cedrus*, *Pinus*, *Keteleeria* and *Tsuga* (Pinaceae)
Simple polyembryony (S)	*Picea*, *Larix*, *Pseudolarix* and *Pseudotsuga* (Pinaceae)
Cleavage polyembryony (C)	*Torreya nucifera* and possibly *Taxus cuspidate* (Taxaceae)
No polyembryony (NP)	*Torreya taxifolia* (Taxaceae)

observed to occur with the same frequency as polyzygotic embryos (Burdon and Zabkiewicz 1973). This is also the case for many genera in the Cupressaceae and most Podocarpaceae genera (Roy Chowdhury 1962).

The timing of cleavage polyembryony is another interesting character variant. With *Pinus*, development of cleavage polyembryony takes place at an early stage of suspensor elongation but in the other genera in the Pinaceae, development is delayed until later stages of embryo development (Konar and Oberoi 1969).

7.3.2 Pattern of Simple Polyembryony (S)

Picea, *Larix*, *Pseudolarix* and *Pseudotsuga* (Pinaceae) have no cleavage polyembryony (Dogra 1967; Roy Chowdhury 1962; Konor and Oberoi 1969). Cleavage polyembryony also rarely occurs in *Abies* (Pinaceae). Within the Cupressaceae, *Thuja* and *Arthrotaxis* have only simple polyembrony (Singh and Oberoi 1962; Roy Chowdhury 1962). *Araucaria* and *Agathis* (Araucariaceae), *Cephalotaxus* and several *Podocarpus* species also have only simple polyembryony (Roy Chowdhury 1962).

7.3.3 Pattern of Cleavage Polyembryony (C)

This pattern is rare. It is only known to occur in the Taxaceae for one or two species, possibly *Torreya nucifera* and possibly *Taxus cuspidata* (Roy Chowdhury 1962; Konar and Oberoi 1969). It is not clear if multiple archegonia are present but simply not fertilized or whether a single archegonium is present at the time of fertilization.

7.3.4 Pattern of No Polyembryony (NP)

This pattern is the most exceptional of the four. Only *Torreya taxifolia* (Taxaceae) lacks either form of polyembryony (Coulter and Land 1905). Here, the ovule has a single archegonium. Upon its fertilization, a single embryo develops.

7.4 From Many to One: The Story of a Single Dominant Embryo

Multiple embryos are commonly observed during early seed development yet a mature seed contains a single embryo. When does the reduction in embryo numbers occur? The answer is taxon-dependent (Singh 1978, pp. 209–211).

Cleavage embryogeny is accompanied by intense competition prior to seed maturity. Consider the case of multiple embryos, originating from both simple and cleavage polyembryony. In *Pinus* spp., these will undergo intense early competition (Dogra 1967).

The competitive advantage is conferred by the embryo's position within the ovule, rather than its genotype. This was reported for *Pseudotsuga menziesii* where the developing embryo closest to the female gametophyte's corrosion cavity became the dominant embryo (Orr-Ewing 1957). In *Pinus taeda*, the competition rapidly comes to a close when the dominant embryo is determined; all other embryos degenerate (Skinner 1992). This happens because the suspensor network actively suppresses subordinate embryos (Cairney et al. 2000). Two waves of programmed cell death are observed at this time; one of these eliminates the losing embryos (Filonova et al. 2002). Polyembryony is generally transient, not persistent, during seed maturation and the timing of death varies among taxa (Singh 1978, pp. 206–209).

This reduction in embryo numbers has been considered to be a form of reproductive compensation (Porcher and Lande 2005) or a form of brood cannibalism (Haig 1992) but these are testable hypotheses rather than fact. Other causes of embryo death, i.e. adult sporophyte-induced or perhaps gametophyte-induced death to multiple embryos has not been ruled out. Transient polyembryony (Singh 1978, p. 209) is the norm of the Pinaceae. Persistent polyembryony is the rare exception (Dogra 1967; Singh 1978, pp. 209–211).

7.5 Stage of Embryo Development

Embryo development for the Pinaceae and most conifers proceeds through three major stages: proembryo, early embryogeny and late embryogeny (Gifford and Foster 1989; Grob et al. 1999). These three stages are shown for *Pinus taeda* in Photo 7.1.

Proembryo formation: Characterized by a free nuclear state. The proembryo expands within its archegonium towards the chalazal end of the ovule, eventually breaking through the archegonial jacket and elongating into a corrosion cavity formed within the female gametophyte (Photo 7.1).

Early embryogeny: Begins when the proembryo and its suspensor mass elongates outside of its archegonium (Photo 7.1), pushes into the female gametophyte and finally ends with formation of a root generative meristem (Singh 1978, p. 188).

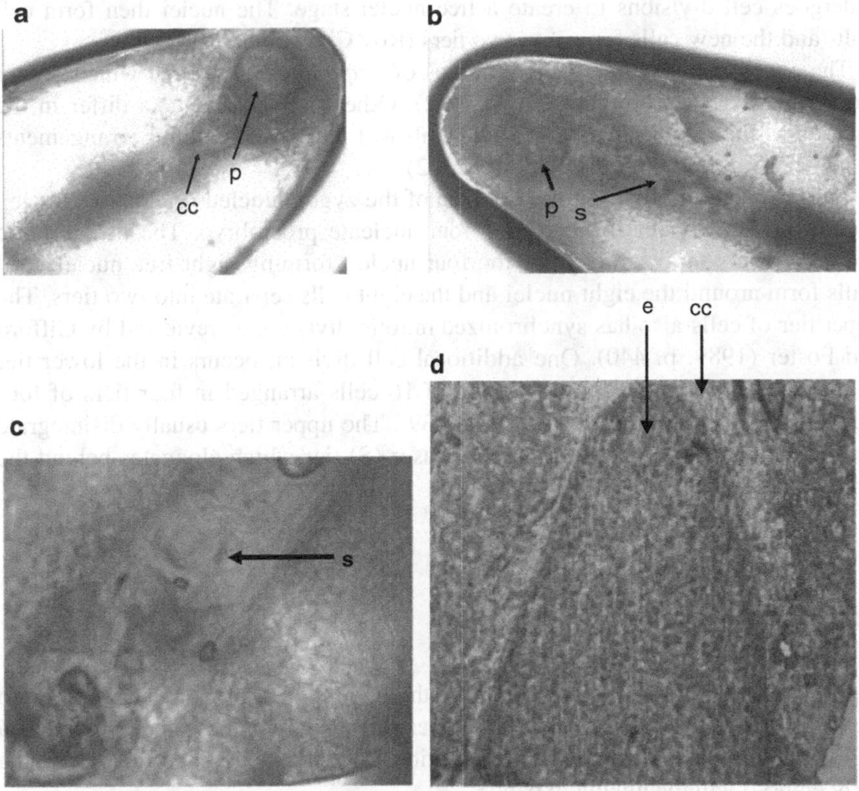

Photo 7.1 Stages of *Pinus taeda* development: (**a**) proembryo (p) stage is shown here as the female gametophyte's corrosion cavity (cc) is forming (arrows) (**b**) the proembryo (p) stage has now formed the suspensor (s). The proembryo has just started to form a suspensor. The corrosion cavity is forming inside the female gametophyte, shown here as the dark V-shaped indentation within the female gametophyte (arrow). (**c**) The proembryo's suspensor (s) network coils and buckles as it pushes into the corrosion cavity, signaling the start of early embryogeny. The proembryo and the suspensor are indicated by the arrow. (**d**) A single dominant selfed *Pinus taeda* embryo (e) is extending into the corrosion cavity (cc) of its female gametophyte during late embryogeny (Adapted from Williams C.G. 2008. Selfed embryo death in *Pinus taeda*: a phenotypic profile. New Phytologist 178:210–222. Copy-right permission granted.)

Late embryogeny: Characterized by the development of polar meristems for primary root and shoot (Singh 1978, p. 188). The shoot meristem is directed away from the micropyle towards the chalazal end. Its radicle (root) end grows towards to the micropylar end of the seed (Photo 7.1).

7.5.1 Proembryo Formation

Proembryo development begins with the fertilized zygote and ends with the elongation of a suspensor (Doyle 1963). As a general rule, the zygote nucleus first undergoes cell divisions to create a free nuclei stage. The nuclei then form cell walls and the new cells form into two tiers (Roy Chowdhury 1962).

The exception is found in the embryos of *Sequoia sempervirens* which do not form free nuclei (Looby and Doyle 1942). Otherwise, conifer taxa differ in the duration of the free nuclear stage, the number of nuclei and cell tier arrangements (Singh 1978, p. 192; Roy Chowdhury 1962).

In *Pinus* spp. the first mitotic division of the zygote nucleus yields two nuclei, each of which divides to become a four-nucleate proembryo. The next mitotic division occurs in synchrony for the four nuclei, forming eight free nuclei. Cell walls form around the eight nuclei and the eight cells separate into two tiers. The upper tier of cells also has synchronized mitotic division, as reviewed by Gifford and Foster (1989, p. 440). One additional cell division occurs in the lower tier so that the proembryo now has a total of 16 cells arranged in four tiers of four cells each (Gifford and Foster 1989, p. 439). The upper tiers usually disintegrate but the outward two tiers form the suspensor (S) tier which elongates behind the embryonal (E) tier.

7.5.2 Early Embryogeny

Vigorous extension forces the proembryo through the wall of its archegonium into the female gametophyte. By now, the upper part of the female gametophyte has a hollow space, the V-shaped corrosion cavity (Buchholz 1918) and this where the early embryo will eventually expand.

The embryo's first embryonal suspensor cells (Es^1) form. The growing suspensor mass buckles as it differentiates cell lineages from which develop independent, genetically identical embryos. As the suspensors elongate, the corrosion cavity becomes too confining. Embryonal suspensors coil and buckle within the cavity (Photo 7.1). As these grow, additional cell divisions in the apical tier form more embryonal suspensor cells (Es^2 and Es^3) forming a larger suspensor mass. The suspensor mass has many roles at this stage (Box 7.1), providing structural support, synthesizing growth regulators and storage products (Ciavatta et al. 2002).

In *Pinus*, the suspensor develops an embryonal mass which in turn develops into cleavage embryos. As the suspensors elongates, cleavage embryos are pushed into a chalazal direction towards the rich nutrient reserves of the female megagametophyte. Competition intensifies among early embryos until a dominant embryo emerges.

In *Picea* spp. and other taxa which lack cleavage polyembryony, the final tier then divides to form a multicellular mass and this mass differentiates into a single embryo (Roy Chowdhury 1962). Seasonal progression of these events in *Pinus taeda* up to early embryogeny (Table 7.3) has been compared across years at the same location (Skinner 1992; Williams 2008).

Box 7.1 Cloning from suspensors for somatic embryos

Few members of the Pinaceae can be vegetatively propagated on a large scale without technology intervention and *Pinus taeda* is no exception. Cloning technology via somatic embryogenesis was developed over 2 decades ago. This was a major technology breakthrough because somatic embryogenesis allows for unlimited copies of a single genotype. The hitch is that the developing seed, yet to be field tested, is the source of somatic embryos. This means that the decision to propagate must be made in the absence of direct phenotypic measurements on the individual itself.

A developing zygotic embryo is isolated at the suspensor stage then cultured in proliferating cell suspension cultures. Large numbers of somatic embryos can form from the single zygotic embryo in the cell suspension cultures. These somatic embryos can be dessicated for long-term storage then sown into growth media where they germinate as somatic seedlings (or emblings) for plantation forestry.

The immediate value of somatic embryogenesis has been to propagate commodity conifers which could not be cloned on a large scale otherwise. Unlimited clonal propagation has also made genetic transformation of *Pinus taeda* and other members of the Pinaceae feasible on a cost-effective commercial scale (von Arnold et al. 2002). Genetically transformed conifers have been controversial given their perennial habit, abundant seed production and long-distance pollen transport (Williams 2006).

Somatic embryos provide a powerful research platform for studying basic of conifer embryo developmental pathways at the physiological, metabolic and DNA levels (von Arnold et al. 2002; Ciavatta et al. 2002).

7.5.3 *Late Embryogeny*

By the third phase of embryogeny, the dominant embryo now has a hypocotyl and a radicle (Grob et al. 1999). A mature pine embryo has a whorl of cotyledons around its shoot apex, short hypocotyls and a primary radicle (Spurr 1949). The female gametophyte consumes the nucellus as source of nourishment from the sporophyte so that at the end of seed maturation, the nucellus persists only as a papery cap of dry tissue at the micropylar end.

The embryo, female gametophyte and the dwindling nucellus are now protected by a stony seed coat, derived from the diploid integument. The stony layer of the integument becomes the hard resistant shell which encloses and protects female gametophyte and embryo against mechanical forces (Gifford and Foster 1989, p. 334).

Table 7.3 Cone, ovule and embryo developmental progression for *Pinus taeda* at Summerville, South Carolina USA across different years as reported by Skinner (1992). Development tends to be synchronous within a single strobilus, conelet and maturing cone

Development	Description	Dates within a single year
Free nuclear stage	Megaspore membrane surrounds the liquid center of the gametophyte	February–May
Cell wall formation	Nuclei in the gametophyte form a monolayer against the megaspore cell wall, separating female gametophyte from the megaspore	May 12–18
Archegonial initials, archegonia	Archegonial initials arise from one to four cells that did not undergo division at the micropylar end. The central cell forms the archegonial jacket	May 19–25
Pollen tube growth	Pollen tube advances	May 19–25
Gametophyte growth	Primary neck cells divide, central cell volume increases and its nucleus is adjacent to the neck cells	May 26–June 1
Fertilization	Mature central canal forms the egg cell and the ventral canal. The neck cells recess and the pollen tube penetrates	June 9–15
Zygote	Stage 1: zygote undergoes mitotic divisions	June 9–15
Free nuclear proembryo	Stage 2: two mitotic divisions create four nuclei that migrate to the opposite end of the archegonia	June 16–22
Proembryo	Stages 3–4: mitotic divisions then subsequent cell wall formation form two tiers	June 16–22
Suspensor	Stage 5: four tiers of cells now form a suspensor network	June 16–22
Cleavage polyembryony	Embryonal tiers split into four embryonal units	June 16–22
Initial embryo proper	Stage 6: embryonal suspensor breaks through the wall of archegonial jacket	June 23–29
Embryo proper at ¼ length	Stage 7: embryo grows into corrosion cavity	June 23–29
Embryo proper at ½ length	Stage 8: corrosion cavity at 60% A dominant embryo can be seen	June 30–July 6
Full-length embryo	Stage 9: embryo is globular and organized	August 25–31
Mature seed		September–October

Remnants of the suspensor form a dry cap which attaches to the root tip of the mature embryo (Roy Chowdhury 1962). Seed dormancy after dispersal is common for many conifers. The female gametophyte does not completely succumb to programmed cell death until the embryo reaches early germination (He and Kermode 2003).

7.6 Seed Dispersal

In *Pinus taeda*, the maturing cone dries and its scales open, releasing mature conifer seeds. Like many conifer seeds, these are winged although the wings are not truly part of the seed. The seed wing, derived from diploid tissues, is formed from the ovuliferous cone scale (Gifford and Foster 1989, pp. 442–443).

Seed yield is abundant for mature trees. Annual production of 74,000 sound seeds per hectare is considered low for a mature *Pinus taeda* forest (Baker and Langdon 1990); sound seed yields exceeding 1.6 million have been reported for old-growth *Pinus taeda* forests within the natural range of the species (Cain and Shelton 2001).

A sympatric relative, *Pinus serotina*, and other *Pinus* spp. worldwide have another pattern: serotiny. Seeds are mature but some do not disperse from the cone until fire causes the release of the seeds from closed cones (Tapias et al. 2001).

Secondary dispersal of pine seeds is also an interesting variant. A number of *Pinus* spp. species have developed mutualistic relationship with corvids (jays), other birds and a number of other vertebrates.

7.6.1 Windborne Dispersal

This dispersal method, prevalent among conifers, has received much experimental and theoretical attention by far. Like pollen, windborne seed dispersal is bimodal because two processes are operative: local neighborhood dispersal (LND) and long-distance dispersal (LDD) (Hengeveld 1989; Nathan et al. 2002). Predicted long-distance seed dispersal distances ranged from 11.9 to 33.7 km from source within *Pinus taeda* plantations in North Carolina (Williams et al. 2006).

While this prediction is consistent with an anecdotal report of *P. radiata* seed dispersal at distances of 8–25 km in a Southern Hemisphere country (Richardson et al. 1994), these two estimates are not directed comparable. The anecdote refers to a case of exotic introduction; many *Pinus* spp. species, indigenous to the Northern Hemisphere, have been repeatedly introduced into South Africa, Australia and other Southern Hemisphere countries for over a century. A few of these exotic introductions disperse seed so effectively by wind, fire or animals that they are now considered invasive to treeless ecosystems in southern Africa (Richardson et al. 1994; Higgins and Richardson 1997; Richardson and Higgins 1998). It is perplexing why seed dispersal distances across a treeless landscape do not exceed those for a plantation setting.

In either case, these long-distance (LDD) estimates of seed dispersal far exceed average seed dispersal distances of 60–90 m reported for *Pinus taeda* (Baker and Langdon 1990). Average dispersal distances are not a correct measure in light of more than one dispersal processes. Besides, using an average is biased in favor of local neighborhood dispersal distances.

7.6.2 Bird-Mediated Dispersal

Eight pine species in the subgenus *Strobus* are known to be dispersed by seed-storing birds, particularly nutcracker jays in the bird family Corvidae. In North America, Clark's nutcracker (*Nucifraga columbiana*) stores seeds from *Pinus edulis*, *P. monophylla*, *P. flexilis* and *P. albicaulis* while its Eurasian relative *N. caryocatactes* caches seeds from *P. cembra*, *P. siberica*, *P. pumila* and *P. koraiensis* in the Carpathian Mountains, Siberia, Mongolia, Korea, China and Japan (Tomback 1978; Tomback and Linhart 1990).

Birds are known to transport pine seed as far as 12–22 km from source (Tomback and Linhart 1990) and the seed shadow is more unpredictable than species which are windborne. A single bird can cache 32,000 seeds (Tomback 1982) which represents three to five times of its energy requirements (Lanner 1982). The birds store the seeds in soil at depths of 2–3 cm under conditions which favor germination if the bird does not retrieve its cache (Tomback 1982; Krakowski et al. 2003). Seeds retain roughly 56% of their viability in the first year but can remain as high as 24% by the fourth year in the cache (Tomback 1982). Such unusual dispersal patterns by birds and other mammals shape the genetic variation of the species (Krakowski et al. 2003) (Box 7.2).

Box 7.2 Edible pine seeds prized as a food source for humans

The term "pine nuts" is not correct. This is because pine, as a gymnosperm, lacks the carpel of a true nut so the correct term is edible pine seeds.

Present-day supplies of edible pine seeds now come from 13 species (Table 7.4). The demand for edible pine seeds has grown in the United States to an annual market of $100 million. This soaring demand was once met with pinyon pines and other indigenous U.S. species but today U.S. markets are supplied mostly from pine species in China and Siberia.

Humans have a long culinary history for the edible pine seed. In Europe, *Pinus pinea* seeds were associated with Neanderthal groups in the Iberian Pennisula as far back as 18,000–49,000 BP (Vendramin et al. 2008). In North America, the story is similar although more recent: prehistoric records from Danger Cave in western Utah date back to 7,500 BP where pinyon pine (*Pinus monophylla*) and limber pine (*P. flexilis*) seeds have been found in human coprolites (Rhode and Madsen 1998). Likewise, edible stone pine seeds were traded by the Phoenicians and later by the Romans to such an extent that the present-day species' range now corresponds to these trading routes along the Mediterranean coast (Vendramin et al. 2008). Edible pine seeds were so prized by the Romans that they were left as sacrificial food at altars all across Europe. Pine seeds, figs and dates were left at alters dedicated to Isis, Cybele and Dionysos and other deities over 2,000 years ago (Zach 2002). Edible pine seeds are intertwined with human history in Europe and the Americas across millennia.

Table 7.4 Pine species valued for present-day human unshelled pine seed consumption

Species Range	Species
Mediterranean	*Pinus cembra*
	Pinus pinea
Western North America	*Pinus monophylla*
	Pinus edulis
	Pinus lambertiana
	Pinus torreyana
Mexico and Central America	*Pinus cembroides*
Eastern Asia	*Pinus koraiensis*
	Pinus armandi
	Pinus gerardiana
	Pinus bungeana
Northern Eurasia	*Pinus sibirica*
	Pinus pumila

7.7 Closing

By opening an edible pine seed, one can see both haploid and diploid phases of the diplohaplontic life cycle. At one time or the other, the developing seed has all parts of the complex life cycle. This edible pine seed is a visual reminder that the that female gametophyte has long been assumed to be a passive participant in embryo development. But is this truly the case?

A surprising degree of variation can be seen over the course of seed development. Conifers and perhaps other gymnosperms have two or more patterns of organelle inheritance – and four patterns of polyembryony. Polyembryony tends to be a transient state but some interesting exceptions do occur. Adaptations for seed dispersal attest to the colonizing ability of these Mesozoic relics. Most conifer seeds are wind dispersed. But equally important forms of dispersal are serotiny, bird-mediated seed caches and secondary dispersal by other vertebrates. Humans are no exception; edible pine seeds have been a food of choice for travelers, traders and cave-dwellers. So much can be read into the sculpted shape of the embryo and its hollow channel inside its female gametophyte.

References

Baker, J. and O. Langdon. 1990. *Pinus taeda* L, Loblolly Pine. In: *Silvics of North America*, Vol. 1. *Conifers*. Agriculture Handbook 654. Forest Service, United States Department of Agriculture, Washington, DC, pp. 497–512.

Buchholz, J. 1918. Suspensor and early embryo of *Pinus*. Botanical Gazette 66: 185–228.

Burdon, R. and J. Zabkiewicz. 1973. Identical and non-identical seedling twins in *Pinus radiata*. Canadian Journal of Botany 51: 2001–2004.

Cain, M. and M. Shelton. 2001. Twenty years of natural loblolly and shortleaf pine seed production on the Crossett Experimental Forest in southeastern Arkansas. Southern Journal of Applied Forestry 25: 40–45.

Cairney, J., N. Xu, et al. 2000. Transcript profiling: a tool to assess the development of conifer embryos. In Vitro Cell Dev Biol 36: 155–162.

Ciavatta, V., U. Egertsdotter, et al. 2002. A promoter from the loblolly pine PtNIP1 gene directs expression in an early embryogenesis and suspensor-specific fashion. Planta 215: 694–698.

Coulter, J. and W. Land 1905. Gametophytes and embryo of *Torreya taxifolia*. Botanical Gazette 39: 161–178.

Doak, C. 1932. Multiple male cells in *Cupressus arizonica*. Botanical Gazette 94: 168–182.

Dogra, P. 1967. Seed sterility and disturbances in embryogeny in conifers with particular reference to seed testing and tree breeding in Pinaceae. Studia Forestalia Suecica 45: 1–97.

Doyle, J. 1963. Proembryogeny in *Pinus* in relation to that in other conifers – survey. Scientific Proceedings of the Royal Dublin Society Section B 62: 181–216.

Filonova, L., S. von Arnold, et al. 2002. Programmed cell death eliminates all but one embryo in a polyembryonic plant seed. Cell Death and Differentiation 9: 1057–1062.

Gifford, E. and A. Foster. 1989. *Morphology and Evolution of Vascular Plants*. W.H. Freeman, New York.

Grob, J., W. Carlson, et al. 1999. Dimensional model of zygotic Douglas-fir embryo development. International Journal of Plant Sciences 160: 653–662.

Guo, F.-L., S.-Y. Hu, et al. 1999. Cytological mechanism of cytoplasmic inheritance in *Pinus tabulaeformis*: II. Transmission of male and female organelles during fertilization and proembryo development. Acta Botanica Sinica 42: 341–352.

Haig, D. 1992. Brood reduction in gymnosperms. Editors: M. Elgar and B. Crespi. In: *Cannibalism: Ecology and Evolution Among Diverse Taxa*. Oxford University Press, Oxford, pp. 63–84.

He, X. and A. Kermode. 2003 . Nuclease activities and DNA fragmentation during programmed cell death of megagametophyte cells of white spruce (*Picea glauca*) seeds. Plant Molecular Biology 51: 509–521.

Hengeveld, R. 1989. *Dynamics of Biological Invasions*. Chapman & Hall, London, 160 p.

Higgins, S. and D. Richardson. 1997. Pine invasions in the Southern Hemisphere: modeling interactions between organism, environment and disturbance. Plant Ecology 135: 79–93.

Hiratsuka, R., Y. Yamada, et al. 2002. Programmed cell death of *Pinus* nucellus in response to pollen tube penetration. Journal of Plant Research 115: 141–148.

Kaur, D. and S. Bhatnagar. 1984. Fertilization and formation of neocytoplasm in *Agathis robusta*. Phytomorphology 34: 56–60.

Konar, R. and Y. Oberoi. 1969. Studies on the morphology and embryology of *Podocarpus gracilior* Pilger. Beiträge zur Biologie der Pflanzen 45: 329–376.

Krakowski, J., S. Aitken, et al. 2003. Inbreeding and conservation in whitebark pine. Conservation Genetics 4: 581–593.

Lam, E. 2004. Controlled cell death, plant survival and development. Nature Reviews Molecular Cell Biology 5: 305–315.

Lanner, R. 1982. Adaptations of whitebark pine for seed dispersal by Clark's Nutcracker. Canadian Journal of Forest Research 12: 391–402.

Looby, W. and J. Doyle. 1942. Fertilization and proembryo formation in *Sequoia*. Proceedings of the Royal Dublin Society 21: 457–476.

McWilliam, J. and F. Mergen. 1958. Cytology of fertilization in *Pinus*. Botanical Gazette 119: 246–249.

Mogensen, H. 1996. The hows and whys of cytoplasmic inheritance in seed plants. American Journal of Botany 83: 383–404.

Nathan, R., G. Katul, et al. 2002. Mechanisms of long-distance dispersal of seeds by wind. Nature 418: 409–413.

Neale, D., K. Marshall, et al. 1991. Inheritance of chloroplast and mitochrondrial DNA in incense cedar (*Calocedrus decurrens* Torr.). Canadian Journal of Forest Research 21: 717–720.

Neale, D. and R. Sederoff. 1989. Paternal inheritance of chloroplast DNA and maternal inheritance of mitochondrial DNA in loblolly pine. Theoretical and Applied Genetics 77: 212–216.

Ohba, K., M. Iwakawa, et al. 1971. Paternal transmission of a plastid anomaly in some reciprocal crosses of sugi, *Cryptomeria japonica* D. Don. Silvae Genetica 20: 101-107.

Orr-Ewing, A. 1957. A cytological study of the effects of self-pollination on *Pseudotsuga menziesii* (Mirb.) Franco. Silvae Genetica 6: 179–185.

Owens, J. and S. Morris. 1990. Cytological basis for cytoplasmic inheritance in *Pseudotsuga menziesii.* I. Pollen tube and archegonial development. American Journal of Botany 77: 433–445.

Pennell, R. and P. Bell. 1988. Insemination of the archegonium and fertilization in *Taxus baccata* L. Journal of Cell Science 89: 551–559.

Porcher, E., and R. Lande. 2005. Reproductive compensation in the evolution of plant mating systems. New Phytologist 166: 673–684.

Rhode, D. and D. Madsen. 1998. Pine nut use in the Early Holocene and beyond: the Danger Cave archaebotanical record. Journal of Archaeological Science 25: 1199–1210.

Richardson, D. and S. Higgins. 1998. Pines as invaders in the Southern Hemisphere. Editor: D. Richardson. In: *Ecology and Biogeography of* Pinus. Cambridge University Press, Cambridge, pp. 450–473.

Richardson, D., P. Williams, et al. 1994. Pine invasions in the Southern Hemisphere: determinants of spread and invadibility. Journal of Biogeography 21: 511–527.

Roy Chowdhury, C. 1962. The embryology of conifers – a review. Phytomorphology 12: 313–338.

Runions, C. and J. Owens. 1999. Sexual reproduction of interior spruce (Pinaceae). I. Pollen germination to archegonial maturation. International Journal of Plant Sciences 160: 631–640.

Singh, H. 1978. *Embryology of Gymnosperms*. Gebrüder Borntraeger, Berlin, 302 p.

Singh, H. and Y. Oberoi. 1962. A contribution to the life history of *Biota orientalis* Endl. Phytomorphology 12: 373–393.

Skinner, D. 1992. Ovule and embryo development, seed production and germination in orchard grown control pollinated loblolly pine (*Pinus taeda* L.) from coastal South Carolina. Department of Biology, Victoria, BC, University of Victoria, 88 pp.

Spurr, A. 1949. Histogenesis and organization of the embryo in *Pinus strobus* L. American Journal of Botany 36: 629–641.

Tapias, R., L. Gil, et al. 2001. Canopy seed banks in Mediterranean pines of south-eastern Spain: a comparison between *Pinus halepensis* Mill., *P. pinaster* Ait., *P. nigra* Arn. and *P. pinea* L. Journal of Ecology 89: 629–638.

Tomback, D. 1982. Dispersal of whitebark pine seeds by Clark's Nutcracker: a mutualism hypothesis. Journal of Animal Ecology 51: 451-467.

Tomback, D. 1978. Dispersal of whitebark pine seeds by Clark's nutcracker: a mutualism hypothesis. Journal of Animal Ecology 51: 451–467.

Tomback, D. and Y. Linhart 1990. The evolution of bird-dispersed pines. Evolutionary Ecology 4: 185–219.

Vendramin, G., B. Fady, et al. 2008. Genetically depauperate but widespread: the case of an emblematic Mediterranean pine. Evolution 62: 680–688.

von Arnold, S., I. Sabala, et al. 2002. Developmental pathways of somatic embryogenesis. Plant Cell, Tissue and Organ Culture 69: 233–249.

Wagner, D., J. Dong, et al. 1991. Paternal leakage of mitochondrial DNA in *Pinus*. Theoretical and Applied Genetics 82: 510–514.

Williams, C., S. LaDeau, et al. 2006. Modeling seed dispersal distances: implications for transgenic *Pinus taeda.* Ecological Applications 16: 117–124.

Williams, C. 2007. Re-thinking the embryo lethal system within the Pinaceae. Canadian Journal of Botany 85: 667–677.

Williams, C. 2008. Selfed embryo death in *Pinus taeda*: a phenotypic profile. New Phytologist 178: 210–222.

Williams, C. 2006. The question of commercializing transgenic conifers. pp. 31–43. Chapter 2. Editor: C.G. Williams. In: *Landscapes, Genomics and Transgenic Conifers.* Springer, Dordrecht, The Netherlands, 270 p.

Zach, B. 2002. Vegetable offerings on the Roman sacrificial site in Mainz Germany - short report on the first results. Vegetation History and Archaebotany 11: 101–106.

Section III
Mating System Dynamics: Form Versus Chance

Plate III A case of *Pinus elliottii* polyembryony from a self-pollinated mating. The photograph of two viable seedlings emerging from a single *Pinus elliottii* seed provides a full explanation of the paternal parentage. What is the genetic model? Conifers, gymnosperms and even early seed plants as a rule develop multiple archegonia (Photograph taken by E. C. Franklin, USDA-Forest Service. Permission granted)

Chapter 8
The Dynamic Wind-Pollinated Mating System

Summary The dynamic wind-pollinated mating system in conifers is more than a random game of pitch and catch; orderly forces work towards maximizing chances of pollen capture while minimizing selfing and interspecific hybridization. Aerodynamics of moving branches, leaves and female strobili favor pollen movement into ovules while more cryptic molecular mechanisms influence paternal parent choice from pollination onward to seed maturity. Outcrossing is the general outcome for most conifers but a few interesting exceptions include mixed mating systems, selfing, hybridization, reproductive sterility and the singular case of paternal apomixis. Self-pollination occurs at moderate rates yet few selfed seed are recovered in some of the Pinaceae; most selfed embryos die before reaching maturity so this is known as the embryo lethal system. Hybrid matings can be blocked by a few pre-zygotic barriers but more often matings between close relatives produce viable, fertile F1 offspring without a change in ploidy. Conifer reproduction is often abundant to the point of nuisance; it is not unusual for a conifer's wind-pollinated mating system to have a genetic footprint extending tens or even hundreds of kilometres from adult trees. At the other extreme, rare cases of reproductive sterility are reported for both the Pinaceae and the Cupressaceae.

Conifers have complex wind-pollinated mating systems. Although they lack colorful flowers and the means to attract bees, birds or butterflies as pollinators, their mating systems are no less intriguing. These wind-pollination systems are closer to an exercise in aerodynamics efficiency between donor and recipient than hit-and-miss capture. Release pollen too soon and the seed will have the same parent as mother and father. Catch pollen originating too far away and the seed might be a hybrid between different species. If synchrony in time and space fails then safeguards select against pollen or the offspring of the unwanted pollen parent. Mate choice is increasingly nonrandom after pollination. But little is known about how selective mechanisms operate in conifers.

Pollen capture occurs within the confines of a local aerodynamic environment created by the female strobilus, needles and branches, all moving with gusting wind currents (Niklas 1982; Niklas 1984). This is known as the turbine model because the female strobilus resembles a turbine. Wind tunnel studies show that the female strobili of *Pinus taeda* generates its own micro-turbulence pattern suited to capture of conspecific pollen (Niklas 1984). Needles create tiny airflow eddies around the female strobilus which act to trap pollen in between flexed cone scales (Niklas 1985).

The aerodynamics of pollination extend beyond the female strobili; in real-time field studies, the branches subtending the female strobilus enhance the local aerodynamic environment in favor of pollen capture. Gusting causes the branches to oscillate such that they sweep in circular arcs. This circular arc alters the inclination angle of the female strobili at the tops of the branches, tilting at a 45-degree angle with strobili tips pointing downwind which favors pollen capture (Niklas 1985). The turbine model is elegant but it is also possible that pollen accumulation occurs only by simple impaction (Cresswell et al. 2007). The aerodynamics of pollination are not yet fully resolved.

Subsequent events leading up to pollen capture are not random either. These are modulated by various recognition systems. One such system exists for the pollination drop which can distinguish a pollen grain from other small particles. Another system signals for pollen tube arrest if non-specific pollen germinates inside the ovule; this has been reported only for pollen from distantly related species or other subgenera (McWilliam 1959; Hagman 1975).

The signal for renewed pollen tube growth just prior to fertilization is also part of a recognition system (Takaso et al. 1996). But perhaps better described is the embryo lethal system. the self-exclusion system operative after zygote formation (Koski 1971). Selfed embryos usually die between the proembryo and late embryogeny stages (Koski 1971). All of these recognition systems, yet to be fully elucidated at the molecular level, show that pollination and mate choice in conifers are orderly events.

8.1 Wind-Pollinated Mating System: Outcrossing or Mixed Selfing/Outcrossing

Nearly all mating systems of conifers are predominantly outcrossing or mixed selfing/outcrossing (Table 8.1). As a general rule, conifers and particularly the Pinaceae, are outcrossing. Predominantly outcrossing species are defined as having less than 5% selfed seed (Brown 1990). Selfing is measured by s where s = 0.05 in this case and its complementary portion (t = 1–s) measures outcrossing. Outcrossing among conifers (Table 8.1) has been well-documented. This can be seen from Table 8.1. which includes a part of the larger meta-analysis study of 52 conifers (O'Connell 2003).

A few exceptions to outcrossing shown in Table 8.1 deserve brief mention. These include *Larix laricina* (Pinaceae) and *Thuja occidentalis* (Cupressaceae) which have population outcrossing rates as low as t = 0.53 (Knowles et al. 1987) and t = 0.51 (Perry and Knowles 1990), respectively. These two species are considered

Table 8.1 Outcrossing rates for a few members of the Pinaceae family (Table modified from O'Connell 2003 and Mitton and Williams 2006)

Species	t	Reference
Picea chihuahuana	0.076	Ledig et al. 1997
Larix laricina	0.729	Knowles et al. 1987
Pinus albicaulis	0.736	Krakowski et al. 2003
Pseudotsuga menziesii	0.752	Stauffer and Adams 1993
Picea glauca	0.730	Innes and Ringius 1990
Abies alba	0.890	Schoeder 1989
Pinus sibirica	0.894	Krutovskii et al. 1995
Pinus taeda	0.994	Friedman and Adams 1985
Picea abies	0.956	Morgante et al. 1991
Pinus sylvestris	0.940	Muona and Harju 1989
Picea mariana	0.924	Boyle and Morgenstern 1986
Pinus koraiensis	0.974	Krutovskii et al. 1995
Pinus ponderosa	0.960	Mitton et al. 1977, 1981
Pinus flexilis	0.980	Schuster and Mitton 2000
Abies procera	0.940	Siegismund and Kjaer 1997
Pinus radiate	0.900	Moran et al. 1980
Picea omorika	1.00	Kuitteinen and Savolainen 1992

to be mixed mating systems rather than predominantly outcrossing mating systems (Brown 1990; Mitton 1992). Such a mixed mating system is a likely explanation for the high degree of population differentiation of fragmented conifer species such as *Cathaya argophylla* (Ge et al. 1998) which is not shown in Table 8.1.

Perhaps the most peculiar case shown in Table 8.1 is *Picea chihuahuana*, a species occurring in small, isolated populations in northwest Mexico. It has population outcrossing rates as low as $t = 0.076$ so it is defined as predominantly selfing (Ledig et al. 1997). It is not yet clear whether this is the lone exception among conifers; all mating systems are not been classified for conifers worldwide.

In the following sections, other exceptions to the outcrossing rule are addressed. These include selfing, interspecific hybridization, reproductive sterility and the unusual case of paternal apomixis (Chapter 2) where unreduced diploid pollen grows into embryos inside its surrogate maternal parent (Pichot et al. 2001).

8.2 Selfing

Selfing, only possible for monoecious conifers, can only be geitonogamous because conifers are monosporangiate. Geitonogamy refers to the case where male and female strobili occur on the same plant but not in a single strobilus (Richards 1997). Conifers do not have autogamy.

8.2.1 Selfing Avoidance

Selfing can be avoided in part by spatial and temporal separation of male and female strobili (Erickson and Adams 1989). Female *Pinus taeda* strobili mostly emerge on the upper branches of the crown, far above the male strobili in older trees (Greenwood 1980); this temporal separation is defined as dichogamy. Younger trees tend to have a prevalence of female or male strobili throughout the crown (Chapter 2). Older *Pinus ponderosa* trees produce more cones and less pollen. Young *Pinus ponderosa* trees are the opposite, serving as pollen donors to older trees loaded with female strobili (Mitton 1992; Mitton and Williams 2006). Proportions of male versus female strobili are also another factor in selfing rates.

Selfing is also avoided if female strobili reach peak receptivity before male strobili on the same tree release pollen (Greenwood 1986). Even so, these are incomplete barriers. Selfing rates fluctuate from year to year with vagaries of weather, wind speeds and age of the tree (Mitton 1992). External factors such as wind speed, wind direction or even stand density determine how much pollen from non-self trees reach receptive female strobili (Mitton 1992; Dyer and Sork 2001). Proportions of selfed pollinations are subject to chance; they vary widely from tree to tree and from year to year. But from pollination onward, mate choice in conifers becomes increasingly nonrandom.

8.2.2 Selfed Embryo Deaths During Development

Direct estimates of selfing are difficult to obtain. Estimated proportions of selfing range from 10% to 25% in *Pinus sylvestris* (Sarvas 1962; Koski 1971). Self-pollination rates are higher than selfed seed recovery (Chapter 9).

The self-incompatibility systems, acting before fertilization, are well characterized for angiosperms but none have not been reported for any conifers yet. Many members of the Pinaceae family exclude selfed embryos via the embryo lethal system. This steep post-fertilization barrier to selfed embryos is attributed to inbreeding depression due to abundant deleterious mutations. More concerted death peaks also occur during embryo development (Williams 2008) but in either case, few viable seeds are recovered from self-pollinations.

8.3 Interspecific Hybridization

Conifers, particularly members of the Pinaceae, do hybridize naturally with sympatric relatives. If two parental species are not closely related then pollen tube arrest or other aberrant signs appear (McWilliam 1959). The degree of incompatibility ranges from slight to strong for many interspecific crosses made between distantly related *Pinus* species or distant related *Picea* species, as shown by Hagman (1975). As one would expect, crosses between soft and hard pines produced the strongest signs of incompatibility between pollen and nucellar tissues; this was described as a pathogen invading a host plant (Hagman 1975).

But other matings between closely related species do produce viable F1 offspring which often prove to be fertile adults. Examples include hybrid complexes for pines documented in the paleobotanical record (Mason 1949) as well as present-day examples found in major centers of species diversity ranging from the southeastern United States (Edwards-Burke et al. 1997), China (Wang et al. 2001; Ma et al. 2006) or the highlands of Mexico (Matos and Schaal 2000). Divergent environmental factors favor stable hybrid complexes; this is the case for the naturally occurring hybrid between Coulter and Jeffrey pine which parallels major physiographic gradients (Zobel 1951). Hybridization is an open-ended, reticulating event; such hybrids not only freely introgress with one another but they can mate with either parental species or even a third sympatric relative.

But the fate of a hybrid event is not always a new species (Table 8.2). The outcome of hybridization in conifers varies widely but it does seem to depend on three factors: (a) hybrid vigor, (b) reproductive isolation and (c) ecological divergence. Rather, a conifer hybrid swarm can reach a state of equilibrium or continue to differentiate into a new species. This has been elegantly shown using a series of phylogenetic reconstructions for two *Pinus* species and their hybrid in Asia (Wang et al. 2001; Ma et al. 2006).

Artificial matings between closely related species tend to produce fertile offspring without a change in ploidy. This was established by early taxonomy studies based on huge, systematic crossability studies (i.e. Righter and Duffield 1951). Successful hybridizations were limited to a subsection or rarely between two closely related subsections. Artificial hybrids have even been made between divergent North American and Asian soft pines within the same subsection and these hybrids become fertile F1 adults (Stone and Duffield 1950). One such hybrid was

Table 8.2 Fate of hybrid complexes for different seed plants. A hybrid complex refers to a group of species in which natural hybridization has occurred. Hybrid complexes are classified according to their fate or stabilization mode of reproduction in the natural hybrids or hybrid progeny (Grant 1981)

Type	Stabilization mode	Examples
Homogamic	Sexual reproducing diploids with normal meiosis	*Pinus* spp. (Pinaceae) *Gilia* spp. (Polemoniaceae) *Eucalyptus* spp. (Myrtaceae)
Clonal complex	Vegetative reproduction	*Opuntia* spp. (Cactaceae)
Agamic	Apomixis; unfertilized seeds	*Citrus* spp. (Rutaceae)
Heterogamic	Permanent translocation heterozygosity; permanent odd polyploidy	*Rosa canina* (Rosaceae) *Oenethera biennis* (Onagraceae)
Polyploid	Sexual reproducing polyploids	*Sanicula* spp. (Umbelliferae) *Asplenium* spp. (Polypodiaceae)

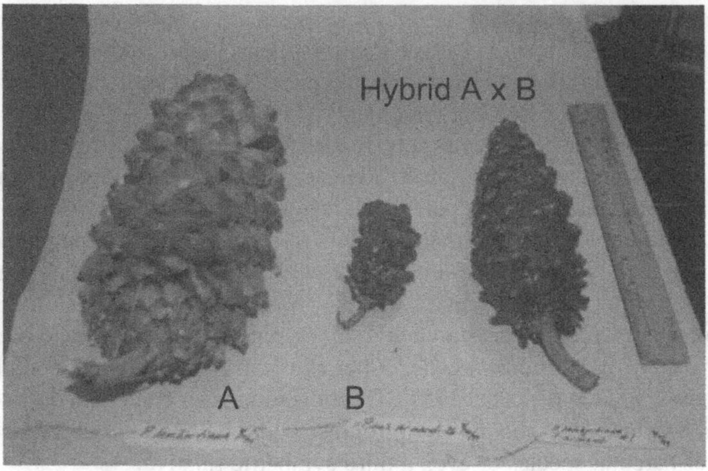

Photo 8.1 A comparison of cone size between a hybrid and its two parental species, a North American soft pine, *Pinus lambertiana* (Parent A) and an Asian soft pine, *Pinus armandii* (Parent B). Their hybrid, planted in the Arnold Arboretum, has a cone size intermediate between the two parental species

known as *Pinus x schwerinii*, came from a cross between Asian soft pine *Pinus wallichiana* x North American soft pine *Pinus strobus* although both parents belong to the same subsection *Strobi* (Price et al. 1998). The hybrid has intermediate reproductive characters such as the cone size comparison (see Photo 8.1).

The fate of hybridization is a useful means of classification (Table 8.2) (Grant 1981). Conifers belong to the homoploid hybrid complex which has the following properties: newly derived species of hybrid origin is diploid or at least undoubled (homoploid) compared to the original parental species, backcrossing to parental species is possible because F1 hybrids have no reproductive isolation created as a direct consequence of hybridization and recombination rates are comparable between hybrid derivatives and their parental species (Grant 1981). Homogamic species complexes are often composed of multiple interfertile species and their hybrids (Grant 1981) suggesting a species complex acts as a single gene pool (Fig. 8.1).

Conifers loosely meet all three conditions. Adult hybrids have stable diploid genomes (Chapter 3), this is the case for the adult hybrid *Pinus x schwerinii* in Photo 8.1 (Williams et al. 2002). This hybrid is also capable of backcrossing to either parent so the second condition is met. Recombination rates are more difficult to quantify. They are similar for two *Larix* species and their hybrid (Sax 1932); however, it is not clear if this condition full met because other F1 hybrids for *Pinus* spp. show lower recombination rates relative to the parental species (Shepherd and Williams 2008). Meiotic abnormalities have also been reported for other *Pinus* F1 hybrids (Saylor and Smith 1966).

What is known about adult F1 hybrids when they reach reproductive onset? In *Pinus* spp., F1 hybrids have a stable diploid genome and stable ploidy (Williams

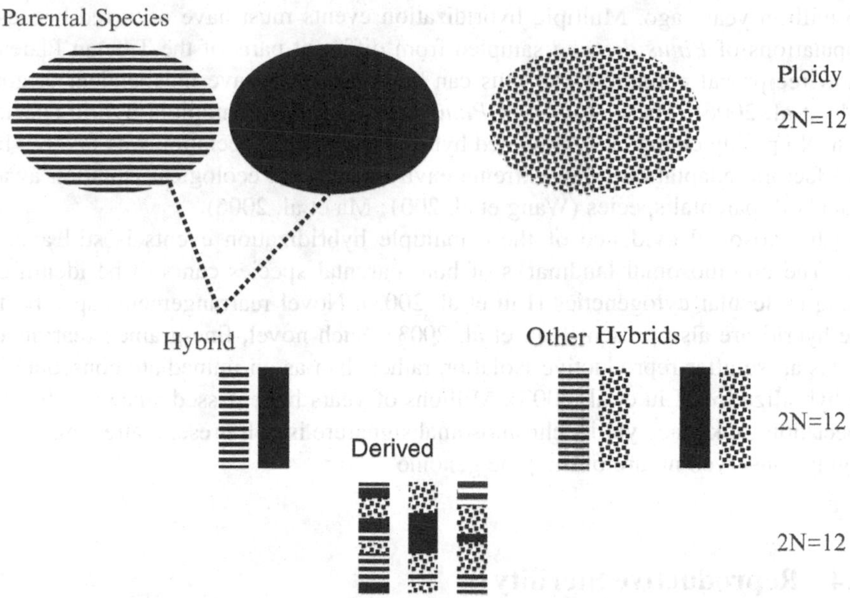

Fig. 8.1 Many conifers mate with sympatric relatives to form homogamic hybrid complexes, as described by Grant (1981) in *Plant Speciation*. Newly derived hybrids are diploid and capable of crossing to their respective parental species. Homogamic hybrid complexes are often composed of multiple, interfertile species and their hybrids

et al. 2002) which adheres to Grant's predictions by the hybrid classification system. No meiotic abnormalities were observed for female meiosis for an adult F1 hybrid soft pine (Sax 1960). Male meiosis for the F1 adult shows more abnormalities such as pollen abortion (Saylor and Smith 1966) and some sterile pollen (Sax 1960). Early stages of male gametophytes sometimes show chromosomal abnormalities and these cannot develop into pollen grains (Saylor and Smith 1966). Still other F1 hybrids have viable pollen grains (Saylor and Smith 1966). No one condition applies to all F1 hybrid adults.

Chromosomal rearrangements such as tiny paracentric inversions could prove useful for reconstructing past hybridization events in conifer species complexes. As shown in many other eukaryotes, paracentric inversions serve the role of recombination suppression in the hybrid genome by preserving large blocks of co-adapted gene complexes. Although hybrid genomes in pines do have unusually high meiotic and genomic stability, the occasional small-scale paracentric inversion can be detected when the F1 hybrids are crossed back to a parental species (Saylor and Smith 1966; Shepherd and Williams 2008).

The best case study of hybridization and its evolutionary consequences is illustrated by *Pinus densata*, a native of the Tibetan Plateau (Wang et al. 2001). This species originated from hybridization between *Pinus tabuliformis* indigenous to northern to central China and *Pinus yunnanensis* which is limited to southwestern China. Hybridization occurred possibly before the uplift of the Tibetan Plateau over

45 million years ago. Multiple hybridization events must have occurred because populations of *Pinus densata* sampled from different parts of the Tibetan Plateau show reciprocal parentage and thus can be assumed to have independent origins (Ma et al. 2006). These studies of *Pinus densata* show homoploid hybrid speciation. No ploidy change accompanied hybrid speciation. Speciation was favored by two factors: adaptation to an extreme environment and ecological isolation away from both parental species (Wang et al. 2001; Ma et al. 2006).

Chromosomal evidence of these multiple hybridization events is still apparent. The chromosomal landmarks of both parental species can still be identified using molecular cytogenetics (Liu et al. 2003). Novel rearrangements specific to the hybrid are also present (Liu et al. 2003). Such novel, fine-grained rearrangements arose after reproductive isolation rather than as an immediate consequence of hybridization (Liu et al. 2003). Millions of years have passed since this hybrid speciation took place yet its chromosomal signature is still present, attesting to the highly conserved nature of the pine genome.

8.4 Reproductive Sterility

The abundant reproduction of the Pinaceae is often taken for granted or even regarded as a nuisance; a single pine tree can produce millions of offspring and countless pollen grains within its long lifespan. But there are rare cases of conifer sterility reported and this holds growing interest to those who want to curtail unwanted seeds and pollen from transgenic or genetically modified conifers (Williams 2006; Williams et al. 2006).

In the first case, a single *Pinus monticola* tree produced male strobili but no pollen (Wilson and Owens 2003). Closer investigation showed that both male and female gametophyte development was arrested soon after meiosis. Pollen grains had poorly developed pollen walls, reduced cytoplasm and did not release. Male strobili aborted before dehiscence. These changes were mediated in part by a malfunction of the tapetal layer. On the female side, meiosis of the megaspore mother cell provided four megaspores yet no female gametophyte developed from any of the megaspores (Wilson and Owens 2003).

In the second case, sterility was the result of aberrant male and female meiosis in *Cryptomeria japonica* (Cupressaceae) (Hosoo et al. 2005). A few maturing pollen grains were produced but these were uneven in size. The female megaspores were also uneven in size. A few megaspores did survive only to die at the archegonium formation stage (Hosoo et al. 2005). Both cases suggest that isolating or inducing tapetal and meiotic mutants are promising research areas for operational sterility in conifers.

A third case is mimicry of reproductive sterility. A bizarre case of insect parasitism causes the mimicry: the chalcid *Megastigmus spermotrophus* infests a female gametophyte inside an unfertilized ovule then induces the female gametophyte to develop normally as though it has been fertilized and has no embryo (von Aderkas et al. 2005a, b). What are the regulatory cues induced by the chalcid?

8.5 Closing

A close look at the conifer mating system shows the precise synchrony between male and female reproductive development through space and time. Although outcrossing is prevalent within a species, selfing, hybridization and reproductive sterility also occur. For monoecious conifers, self-pollination is avoided through spatial or temporal separation of male and female strobili but when it does occur, the selfed embryos are excluded after zygote formation. Conifers are thought to lack the pre-zygotic mechanism of self-incompatibility typical of many angiosperm taxa but this area of research continues using better genomics-based tools. Naturally occurring hybridization can produce fertile adult F1 individuals which are capable of backcrossing to the parental species or even a third species; this raises the difficult question of actual effective population size. Species complexes for conifers are not unusual; they commonly occur in present-day centers of species diversity as well as in the paleobotanical record.

References

Boyle, T. and E. Morgenstern. 1986. Estimates of outcrossing rates in six populations of black spruce in central New Brunswick. Silvae Genetica 35: 102–106.

Brown, A. 1990. Genetic characterization of plant mating systems. Editor: B. Weir. In: *Plant Populations, Genetics, Breeding and Genetic Resources*. Sinauer Associates, Sunderland MA, pp. 145–162.

Cresswell, J., K. Henning, et al. 2007. Conifer ovulate cones accumulate pollen principally by impaction. Proceedings of the National Academy of Sciences USA 104: 18141–18144.

Dyer, R. and V. Sork. 2001. Pollen pool heterogeneity in shortleaf pine, *Pinus echinata* Mill. Molecular Ecology 10: 859–866.

Edwards-Burke, M., J. Hamrick, et al. 1997. Frequency and direction of hybridization in sympatric populations of *Pinus taeda* and *P. echinata* (Pinaceae). American Journal of Botany 84: 879–886.

Erickson, V. and W. Adams. 1989. Mating success in a coastal Douglas-fir seed orchard as affected by distance and floral phenology. Canadian Journal of Forest Research 19: 1248–1255.

Friedman, S. and W. Adams. 1985. Levels of outcrossing in two loblolly pine seed orchards. Silvae Genetica 34: 157–162.

Ge, S., D. Hong, et al. 1998. Population genetic structure and conservation of an endangered conifer, *Cathaya argyrophylla* (Pinaceae). International Journal of Plant Sciences 159: 351–357.

Grant, V. 1981. *Plant Speciation*. Columbia Press, New York.

Greenwood, M. 1980. Reproductive development in loblolly pine. I. The early development of male and female strobili in relation to the long shoot growth behavior. American Journal of Botany 67: 1414–1422.

Greenwood, M. 1986. Gene exchange in loblolly pine: the relation between pollination mechanisms, female receptivity and pollen viability. American Journal of Botany 73: 1433–1451.

Hagman, M. 1975. Incompatibility in forest trees. Proceedings of the Royal Society of London B 188: 313–326.

Hosoo, Y., E. Yoshii, et al. 2005. A histological comparison of the development of pollen and female gametophytes in fertile and sterile *Cryptomeria japonica*. Sexual Plant Reproduction 18: 81–89.

Innes, D. and G. Ringius. 1990. Mating system and genetic structure of two populations of white spruce (*Picea glauca*) in eastern Newfoundland. Canadian Journal Botany 68: 1661–1666.

Knowles, P., G. Furnier, et al. 1987. Significant levels of self-fertilization in natural populations of tamarack. Canadian Journal of Botany 65: 1087–1091.

Koski, V. 1971. Embryonic lethals of *Picea abies* and *Pinus sylvestris*. Communicationes Instituti Forestalia Fennica 75: 1–30.

Krakowski, J., S. Aitken, et al. 2003. Inbreeding and conservation in whitebark pine. Conservation Genetics 4: 581–593.

Krutovskii, K., D. Politov, et al. 1995. Isozyme study of population genetic structure, mating system and phylogenetic relationships of the five stone pine species (subsection *Cembrae*, section Strobi, subgenus Strobus). Editors: P. Baradat, W.T. Adams, G. Muller-Starck. In: *Population Genetics and Conservation of Forest Trees*. Springer, The Netherlands, pp. 270–304.

Kuittinen, H. and O. Savolainen. 1992. *Picea omorika* is a self-fertile but outcrossing conifer. Heredity 68: 183–187.

Ledig, F., V. Jacob-Cervantes, et al. 1997. Recent evolution and divergence among populations of a rare Mexican endemic, Chihuahua spruce, following Holocene climatic warming. Evolution 51: 1815–1827.

Liu, Z.-L., D. Zhang, et al. 2003. Chromosomal localization of 5S and 18S-5.8S-25S ribosomal DNA sites in five Asian pines using fluorescence in situ hybridization. Theoretical and Applied Genetics 106: 198–204.

Ma, X.-F., A. Szmidt, et al. 2006. Genetic structure and evolutionary history of a diploid hybrid pine *Pinus densata* inferred from the nucleotide variation at seven gene loci. Molecular Biology and Evolution 23: 807–816.

Mason, H. 1949. Evidence of the genetic submergence of *Pinus remorata*. Editor: G. Simpson. In: *Genetics, Speciation and Paleontology*. Princeton University Press, Princeton, NJ, 474 p.

Matos, J. and B. Schaal. 2000. Chloroplast evolution in the *Pinus montezumae* complex: a coalescent approach to hybridization. Evolution 54: 1218–1233.

McWilliam, J. 1959. Interspecific incompatibility in *Pinus*. American Journal of Botany 46: 425–433.

Mitton, J. 1992. The dynamic mating system of conifers. New Forests 6: 197–216.

Mitton, J., Y. Linhart, et al. 1977. Observations on the genetic structure and mating system of ponderosa pine in the Colorado Front Range. Theoretical Applied Genetics 7: 5–13.

Mitton, J., Y. Linhart, et al. 1981. Estimation of outcrossing in ponderosa pine, *Pinus ponderosa* Laws., from patterns of segregation of protein polymorphisms and from frequencies of albino seedlings. Silvae Genetica 30: 117–121.

Mitton, J. and C. Williams. 2006. Gene flow in conifers. pp. 147–168. Chapter 9. Editor: C.G. Williams. In: *Landscapes, Genomics and Transgenic Conifers*. Springer, Dordrecht, The Netherlands, 270 p.

Moran, G., J. Bell, et al. 1980. The genetic structure and levels of inbreeding in a *Pinus radiata* seed orchard. Silvae Genetica 29: 190–193.

Morgante, M., G. Vendramin, et al. 1991. Effects of stand density on outcrossing rate in Norway spruce (*Picea abies*) populations. Canadian Journal Botany 69: 2704–2708.

Muona, O. and A. Harju. 1989. Effective population sizes, genetic variability and mating system in a natural stands and seed orchards of Pinus sylvestris. Silvae Genetica 38: 221–228.

Niklas, K. 1982. Simulated and empiric wind pollination patterns of conifer cones. Proceedings National Academy of Sciences U.S.A. 79: 510–514.

Niklas, K. 1984. The motion of windborne pollen grains around conifer ovulate cones: implications on wind pollination. American Journal of Botany 71: 356–374.

Niklas, K. 1985. Wind pollination – a study in chaos. American Scientist 73: 462–470.

O'Connell, L. 2003. The evolution of inbreeding in western red cedar (*Thuja plicata*: Cupressaceae). Department of Forest Sciences, Faculty of Forestry. University of British Columbia, Vancouver, BC, 162 p.

Perry, D. and P. Knowles. 1990. Evidence of high self- fertilization in natural populations of eastern white cedar (*Thuja occidentalis*). Canadian Journal of Botany 68: 663–668.

Pichot, C., M. El-Maataoi, et al. 2001. Surrogate mother for endangered *Cupressus*. Nature 412: 39.

Price, R., A. Liston, et al. 1998. Phylogeny and systematics of *Pinus*. Editor: D. Richardson. In: *Ecology and Biogeography of Pinus*. Cambridge University Press, Cambridge UK, pp. 49–68

Richards, A. 1997. *Plant Breeding Systems*. Chapman & Hall, London.

Righter, F. and J. Duffield. 1951. A summary of interspecific crosses in the genus *Pinus* made at the Institute of Forest Genetics. Journal of Heredity 42: 75–80.

Sax, H. 1932. Chromosome pairing in *Larix* species. Journal of the Arnold Arboretum 13: 368–373.

Sax, K. 1960. Meiosis in interspecific pine hybrids. Forest Science 6: 135–138.

Sarvas, R. 1962. Investigations on the flowering and seed crop of *Pinus silvestris*. Communicationes Instituti Forestalis Fennica 53: 1–198.

Saylor, L. and B. Smith. 1966. Meiotic irregularity in species and interspecific hybrids of *Pinus*. American Journal of Botany 53: 453–468.

Schoeder, S. 1989. Outcrossing rates and seed characteristics in damaged natural populations of *Abies alba*. Silvae Genetica 38: 185–189.

Schuster, W. and J. Mitton. 2000. Paternity and gene dispersal in limber pine (*Pinus flexilis* James). Heredity 84: 348–361.

Shepherd, M. and C. Williams. 2008. Comparative mapping among subsection Australes (genus *Pinus*, family Pinaceae). Genome 51: 320–331.

Siegismund, H. and E. Kjaer. 1997. Outcrossing rates in two stands of noble fir (*Abies procera* Rehd.) in Denmark. Silvae Genetica 46: 144–146.

Stauffer, A. and W. Adams. 1993. Allozyme variation and mating system of three Douglas-fir stands in Switzerland. Silvae Genetica 42: 254–258.

Stone, E. and J. Duffield. 1950. Hybrids of sugar pine by embryo culture. Journal of Forestry 48: 200–201.

Takaso, T., P. von Aderkas, et al. 1996. Prefertilization events in ovules of *Pseudotsuga*: ovular secretion and its influence on pollen tubes. Canadian Journal of Botany 74: 1214–1219.

von Aderkas, P., G. Rouault, et al. 2005a. Multinucleate storage cells in Douglas-fir (*Pseudotsuga menziesii* (Mirbel) Franco) and the effect of seed parasitism by the chalcid *Megastigmus spermotrophus* Wachtl. Heredity 94: 616–622.

von Aderkas, P., G. Rouault, et al. 2005b. Seed parasitism redirects ovule development in Douglas fir. Proceeding of the Royal Society B 272: 1491–1496.

Wang, X.-R., A. Szmidt, et al. 2001. Genetic composition and diploid speciation of a high mountain pine, *Pinus densata*, native to the Tibetan plateau. Genetics 159: 337–346.

Williams, C. 2006. The question of commercializing transgenic conifers. pp. 31–43. Editor: C.G. Williams. In: *Landscapes, Genomics and Transgenic Conifers*. Springer, Dordrecht, The Netherlands, 270 p.

Williams, C. 2008. Selfed embryo death in *Pinus taeda*: a phenotypic profile. New Phytologist 178: 210–222.

Williams, C., K. Joyner, et al. 2002. Genomic consequences of interspecific *Pinus* spp. hybridisation. Biological Journal of the Linnean Socirty 75: 503–508.

Williams, C., S. LaDeau, et al. 2006. Modeling seed dispersal distances: implications for transgenic *Pinus taeda*. Ecological Applications 16: 117–124.

Wilson, V. and J. Owens. 2003. Histology of sterile male and female cones in *Pinus monticola* (western white pine). Sexual Plant Reproduction 15: 301–310.

Zobel, B. 1951. The natural hybrid between Coulter and Jeffrey pines. Evolution 5: 405–413.

Chapter 9
The Embryo Lethal System

Summary Outcrossing, wind-pollinated members of the Pinaceae have high self-pollination rates yet produce few selfed seedlings. Avoiding self-pollen capture is incomplete so how are self-pollinated ovules or seeds selectively eliminated? Barriers to selfing have long been considered to be either competition via simple polyembryony and death to selfed embryos during seed maturation. Experimental results show that simple polyembrony is a weak barrier against selfed embryos. By far, the most effective barrier is the enigmatic mechanism(s) that cause recognition and death to selfed embryos. A survey shows that extreme inbreeding depression occurs in some species but not in others so this is not a feature of conifers as a group. Only five of the 11 genera within the Pinaceae (*Abies*, *Larix*, *Picea*, *Pinus* and *Pseudotsuga*) have been well-characterized with respect to self-pollinated embryo deaths. Molecular dissection methods have been used to infer severity and distribution of lethal factors; to date, most are semi-lethal rather than fully lethal. These range from partially dominant to overdominant or perhaps balanced lethals.

Some selfed embryos die at all stages of seed development but a second death pattern has been detected in some *Pinus* and *Picea* spp species: a large proportion of selfed embryo deaths peak during early embryogeny. Are these dual death patterns present in other genera and if so, what genetic models might account for them? This chapter is a case study which integrates not only what was introduced in previous chapters but also shows how knowledge of the conifer mating system contributes to the broader understanding of eukaryotic systems.

With a single photograph, Plate III shows the full genetic model. The maternal parent must be a heterozygous carrier for the deleterious albino recessive allele. At least two archegonia were present in the ovule because one germinant is albino and the other is wild type. Each archegonium was pollinated with a different pollen grain so these are polyzygotic embryos. One pollen grain had the deleterious albino recessive allele and the other pollen grain had the wild type allele. But there is another anomaly which cannot be deduced from looking at the photograph. Even though this was a self-pollinated mating, the embryos survived. What is unusual about that? The upcoming chapter addresses this question.

C.G. Williams, *Conifer Reproductive Biology*,

9.1 Moderate Selfing Rates yet Low Selfed Seed Recovery

It is curious that more selfed individuals not found within populations for the Pinaceae and other monoecious conifers. Part of the answer is that 1) male and female strobili are spatially separated on the same tree and 2) that peak pollen shed rarely coincides with female strobilus receptivity. But these are incomplete barriers.

Self-pollination rates can be quite high in the middle of the *Pinus taeda* crown where female and male strobili often overlap. They are reported to reach as high as 34% yet only 5% of viable seed is recovered as self-fertilized seeds (Franklin 1969). This is also the case for self-pollination rates for *Pseudotsuga menziesii* which reach as high 40 to 60% without increasing the proportion of selfed seedlings in a population (Sorensen 1982). So how to explain the high proportion of empty selfed seed shown in Fig. 9.1?

Self-pollination (or selfing) occurs at moderate rates yet the selfed seed is not recovered. This means that the selfed embryos must be dying between fertilization and seed maturity. The interval between pollination and fertilization has be ruled out. Self-pollinated ovules do not undergo any morphological changes from pollination to zygote formation, as reported for several members of the Pinaceae: *Pseudotsuga menziesii* (Orr-Ewing 1957), *Pinus peuce* (Hagman and Mikkola 1963), *Picea glauca* (Mergen et al. 1965), *Picea abies* and *Pinus sylvestris* (Koski 1971). Pre-fertilization barriers are notably absent (Sarvas 1962) and none have been detected experimentally using molecular markers and viability assays so far (Kuang et al. 1999; Williams et al. 2001; Williams 2008).

Thus it is clear that loss of self-pollinated ovules is occurring between fertilization and seed maturation. Selfed death might be occurring as a consequence of embryo competition among polyzygotic embryos. The more likely barrier is severe inbreeding depression, also known as the embryo lethal system (Koski 1971).

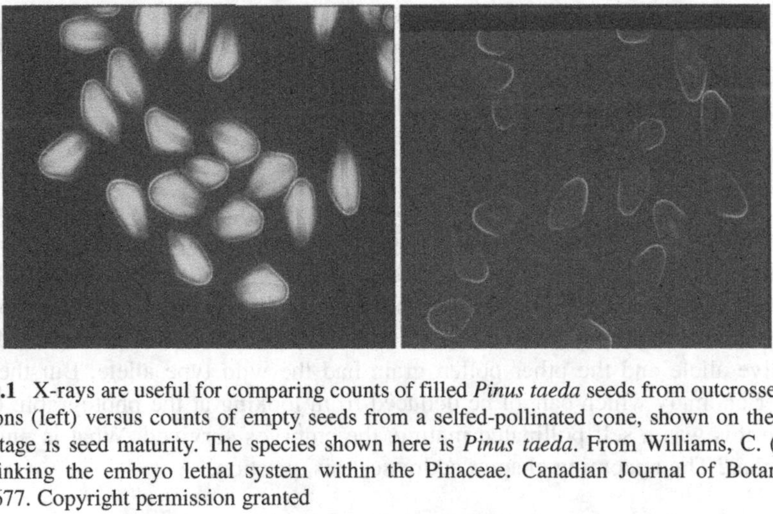

Fig. 9.1 X-rays are useful for comparing counts of filled *Pinus taeda* seeds from outcrossed pollinations (left) versus counts of empty seeds from a selfed-pollinated cone, shown on the right. The stage is seed maturity. The species shown here is *Pinus taeda*. From Williams, C. (2007) Re-thinking the embryo lethal system within the Pinaceae. Canadian Journal of Botany 85: 667–677. Copyright permission granted

9.2 Simple Polyembryony is not a Barrier Against Selfs

It is intuitively appealing to assume that polyembryony acts a selective sieve but this is not so. This idea was first advanced as developmental selection: Buchholz (1922) wrote "…developmental selection is characterized by the fact that it occurs between very minute or embryonic individuals whose struggle is limited to what might be termed an internal environment. It is well-illustrated by the selection resulting from the polyembryony within the developing seeds of conifers and cycads…".

But developmental selection is based on a mistaken observation. The number of embryos inside each archegonium (Buchholz 1918) was overestimated. The rosette tier in the proembryo stage was mistakenly thought divided to form four additional rosette embryos (Skinner 1992). This mistake was further extended to a comparative study of conifer genera (Buchholz 1920) yet still continues to draw commentary (Haig 1992).

The genetic outcome of embryo competition depends on the type of polyembryony. Only selection among polyzygotic embryos (simple polyembryony) has genetic consequences. Cleavage embryos are genetically identical so the dominant embryo will have the same genotype as the losing embryos (Gifford and Foster 1989 p. 439) so for this type, selecting one genotype over another offer no selective advantage. Competition between selfs and nonselfs has to be limited to simple polyembryony.

Even so, simple polyembrony does not account for the differential deaths to selfed seeds (e.g. Sorensen 1982; O'Connell and Ritland 2005). Three sources of additional proof can be offered to these studies. First, competitive advantage is conferred by an embryo's position, not by its genotype. This is the case for *Pseudotsuga menziesii* where the embryo closest to the corrosion cavity gains the competitive advantage whether it is self or nonself (Orr-Ewing 1957).

Second, embryo competition may occur early, before any real differences are expressed between selfs and nonselfs. This appears to be the case for *Pinus* spp. where intense competition occurs at the proembryo stage or at the start of early embryogeny (Dogra 1967). The winner of the competition is a single dominant embryo which then elongates into the corrosion cavity (Dogra 1967; Filonova et al. 2002). If it dies, the embryo is not replaced. Such reproductive compensation offers selective advantage in other eukaryotes (Porcher and Lande 2007) but this is not the case for the Pinaceae.

Third, predictive models such as the lethal number or COMB model (Chapter 8) show that selective advantage of nonself over self is so minimal that any difference between them cannot be detected experimentally (Koski 1971). One reason is that that polyzygotic embryos are more closely related than full-sibs (Box 9.1). Recall that selfed and nonselfed embryos in the same female gametophyte are more closely related than full-sibs because they have an identical maternal haplotype. Weak selective advantage can be expected when relatedness between embryos is so higher than full-sibs.

In short, simple polyembryony carries no adaptive significance towards eliminating selfed embryos. It only serves as a modicum of insurance against numerical odds of having no viable offspring (Sorensen 1982). Simple polyembryony is better viewed as a vestige of the highly conserved development of the female gametophyte rather than a barrier against selfed seeds.

Box 9.1 Genetics of simple polyembryony

The genetics of simple polyembryony in a conifer are illustrated here using photograph shown in Plate III. Assaying two linked molecular markers A and B allows separate tracking of male and female haplotypes in the embryo. Assume that molecular marker A has multiple alleles denoted as A_1,A_2,A_n in the population of adult trees and that it is linked to molecular marker B which has multiple alleles denoted as B_1, B_2 or B_n. The two linked marker alleles delineate a chromosomal haplotype. Each parent's diploid genotype is composed of two haplotypes. Each of their haploid gametes has one haplotype each. Assume that a single recessive gene for albinism is located in the haplotype A_1-B_2. The probability that *both* paternal and maternal parent transmit the same A_1-B_2 haplotype is increased if the parents are related.

First consider the outcrossing case. An ovule has two archegonia. Assume two different unrelated pollen parents each fertilize a different archegonium within the same ovule. The female gametophyte's haplotype, given as A_1-B_2, is also shared by both archegonia in the ovule. Assume that one pollen grain has the A_3-B_4 haplotype and the other pollen grain has the A_5-B_6 haplotype. Genotypes for these two polyzygotic embryos are A_1-B_2/A_3-B_4 and A_1-B_2/ A_5-B_6. Here it is shown how the two outcrossed embryos are more genetically similar than full-sibs yet not as similar as identical twins.

Now consider the selfing case. It is the same adult sporophyte in the outcrossing example. Its genotype is composed of two haplotypes: A_1-B_2 and A_7-B_8. After meiotic recombination in both male and female meiosis, the adult transmits four haplotypes to its megaspores and microspores: two parental-type gametes (A_1-B_2 and A_7-B_8) and two recombinant-type gametes (A_1-B_8 and A_7-B_2). After female meiosis, only one of the four megaspores in the tetrad will develop into a female gametophyte. Let us assume that this haploid female gametophyte received the A_1-B_2 haplotype and under the assumption of monospory, this is also the haplotype for both of its egg cells.

Each of the four microspores becomes a pollen grain. From the same adult sporophyte, it also received one of the same four haplotypes. Here too, each pollen grain fertilizes a different archegonium in the same ovule. If one fertilizing pollen grain has the A_1-B_2 haplotype and the second pollen grain has the A_7-B_8 haplotype, then the first embryo (which has A_1-B_2 from its maternal gamete and A_1-B_2 from its paternal gamete) will receive two copies of the recessive albino mutation and the second embryo (A_1-B_2 from its maternal gamete and A_7-B_8 from its paternal gamete) will be normal or wild-type. But contrary to what normally occurs, the two embryos persist past seed germination.

9.3 Measures of Selfed Embryo Death

The only real barrier to selfed seeds to date is some form of extreme inbreeding depression between the zygotic stage and seed maturity. Increased homozygosity results in more empty seed with self-fertilization (F = 0.5) than other, less related matings (Table 9.1). As expected, selfed matings produce fewer filled seeds than sib matings but that the highest proportion of filled seeds comes from outcrossed matings.

For comparative purposes, Table 9.1 is only a dataset composed of number of filled seeds between selfed and nonself cones. For comparative purposes, one can use standardized methods as lethal numbers or lethal equivalents.

9.3.1 Lethal Equivalents

Calculating lethal equivalents is based on the Morton-Crow-Muller (MCM) method (Morton et al. 1956). In mammals and plants alike, the lethal equivalent number is estimated from the slope of a log-linear survivorship equation constructed among multiple generations of inbred matings (Table 9.2). The slope, B (or 2B for zygotes), represents the hidden genetic damage that would be expressed fully only in a complete homozygote which is the case where the inbreeding coefficient (F) reaches 1. The intercept, A, estimates the amount of expressed damage in a random mating population and thus confounds overdominant genetic effects with environmental effects.

The MCM method has been adapted for a single generation of selfing in conifers (Sorensen 1969). Using Sorensen's (1969) method, the frequency of empty seeds after self- and cross-pollination within a single generation is measured to obtain the relative self-fertility (the ratio of full seeds upon selfing to full seeds upon

Table 9.1 Inbreeding depression for *Pinus patula* seed viability can be calculated using data adapted from Williams et al. (1999). A filled seed, shown in Fig. 9.1, is a viable seed. Note that selfed seed viability is the lowest but that other related matings, half- and full-sibs, also have lower seed viability when compared to outcrossed seeds. From Williams et al. (1999) Embryonic lethal load for a neotropical conifer, *Pinus patula* Schiede and Deppe. Journal of Heredity 90: 394–398. Copyright permission granted

Mating	Total seeds per cone	Filled seeds per cone	Filled seed yield (%)
Outcross	37.8 ± 10.3	17.7 ± 6.5	43.9 ± 6.8
Half-sib	32.9 ± 10.6	8.3 ± 1.4	30.7 ± 5.5
Full-sib	38.4 ± 9.9	10.3 ± 1.5	27.6 ± 6.2
Self	35.1 ± 6.7	3.4 ± 1.3	8.4 ± 1.9

Table 9.2 Lethal equivalents (LE) compared among animals and plants, including conifers. Calculations are based on lethal equivalents per zygote (2B). All estimates are solely based on the offspring viability component. Adapted from Williams, C. (2007) Re-thinking the embryo lethal system within the Pinaceae. Canadian Journal of Botany 85: 667–677. Copyright permission granted

Species	LE (2B)	References
Triggerplant (*Stylidium* spp.)	20.0	Burbridge and James 1991
Eld's deer (*Cervus eldi thamin*)	15.1	Ralls et al. 1988
Tamarack (*Larix laricina*)	10.8	Park and Fowler 1982
High-bush blueberry (*Vaccinium corymbosum*)	10.0	Krebs and Hancock 1991
Douglas-fir (*Pseudotsuga mensiezii*)	8.6	Sorensen 1971
Loblolly pine (*Pinus taeda*)	8.5	Franklin 1972
Loblolly pine (*Pinus taeda*)	7.3	Williams et al. 2001
Loblolly pine (*Pinus taeda*)	9.9	Williams 2008
Reindeer (*Rangifer tarandus*)	8.4	Ralls et al. 1988
Patula pine (*Pinus patula*)	7.2	Williams et al. 1999
Scots pine (*Pinus sylvestris*)	6.6	Koski 1973
Speke's gazelle (*Gazella spekei*)	6.2	Ralls et al. 1988
Japanese quail (*Coturnix coturnix japonica*)	3.6	Sittman et al. 1966
Zebra (*Equus burchelli*)	3.1	Ralls et al. 1988
Fruit flies (*Drosophila melanogaster*)	2.3	Lewontin 1974
Humans (*Homo sapiens*)	2.2	Cavalli-Sforza & Bodmer 1971)
Willow (*Salix viminalis*)	1.8	Kang et al. 1994
Zebrafish (*Danio rerio*)	1.4	McCune et al. 2002
Bluefin killifish (*Lucania goodei*)	1.9	McCune et al. 2002
Serbian spruce (*Picea omorika*)	1.9	Koski 1973
Meadow sage (*Salvia pratensis*)	1.3	Ouborg and Treuren 1994
Cuckoo flower (*Lynchis flos-cuculi*)	0.8	Hauser and Loeschcke 1994
Red drummond phlox (*Phlox drummondii*)	0.8	Levin 1991
Monkeyflower (*Mimulus guttatus*)	0.3	Latta and Ritland 1994
Red pine (*Pinus resinosa*)	0.1	Fowler 1965a,b
Coast redwood (*Sequoia sempervirens*)	0.03	Libby et al. 1981
Western red cedar (*Thuja plicata*)	0	Owens et al. 1990

outcrossing). The number of lethal equivalents was estimated as −4 (ln R), where R is the ratio of selfed seed yields to outcrossed seed yields. This method is useful for comparisons across plant and animal species whether or not they occur as unisex or separate-sex organisms (Table 9.2) but it does not account for the effects of simple polyembryony. The meta-analysis of lethal equivalents for conifers compared to other plants and animals, extended from Williams and Savolainen (1996), is shown in Table 9.2.

Life history alone cannot fully account for the range of lethal equivalent numbers among warm-blooded animals or plants shown here. Some conifers do have high lethal equivalents while others, like the notable exceptions of *Pinus resinosa* have none. This suggests that the perennial habit alone cannot account for severe inbreeding depression at the embryo development stages. Angiosperm perennials such as *Vaccinium* spp. and *Stylidium* spp. do have exceptionally high numbers of lethal equivalents yet *Salix viminalis* does not (Table 9.1). Warm-blooded mammals have a range of lethal equivalents: yet note that humans are at the lower end with lethal equivalent numbers closer to fish or fruit flies (Table 9.1).

9.3.2 Lethal Numbers

The second method of estimating lethals, also known as the combinatorial (COMB) method, accounts for simple polyembryony and thus tends to be used for comparisons among conifers.

Based on data from self- versus outcrossed pollinations, the adult sporophyte is assumed to be heterozygous at *n* loci for recessive embryonic lethals. All offspring receiving two copies of a recessive mutant allele die shortly after fertilization (Koski 1971; Bramlett and Popham 1971). The COMB model also assumes independent gene action (Bramlett and Bridgwater 1986) rather than epistasis (Griffin and Lindgren 1985). Like the MCM model, the COMB model assumes selfed embryo death is caused by the increased proportion of recessive homozygotes that accompanies selfing.

Another larger source of bias in the COMB model, other than archegonial number, is non-genetic deaths (Savolainen et al. 1992). This adjustment removes extraneous mortality which upwardly bias estimates of lethals (Lindgren 1975; Savolainen et al. 1992). A comparison was provided with and without these adjustments for *Pinus patula* (Williams et al. 1999). Changing the archegonial number from one to two raised the lethal number from 9.5 to 11.7 for the population but adjusting for non-genetic causes reduced lethal number from 9.5 to 6.2 while holding archegonial number constant (Williams et al. 1999). The importance of archegonial numbers is skewed upwards one archegonium rarely equates to one fertilized embryo; Only half of the archegonia are fertilized (Skinner 1992).

The MCM and COMB methods provide only phenotypic description. They do not provide information about the underlying genetic basis. A lethal equivalent is defined as "a group of mutant genes…they would cause on the average one death, e.g. one lethal mutant, or two mutants each with 50 per cent probability of causing death…" (Morton et al. 1956). Some inferences about the severity and distribution of lethal-causing chromosomal segments can be provided by molecular dissection methods, as discussed later.

9.4 The Embryo Lethal System

Selfed embryos die between proembryo and late embryogeny stage via the embryo lethal system (Koski 1971). This term originally implied a concerted system but with time its meaning has been equated with deleterious alleles segregating at embryo viability loci (i.e. Koski 1971; Namkoong and Bishir 1987; Remington and O'Malley 2000a).

Recent evidence suggests more than one pattern for selfed embryo death; both models appear to be operative at the same time (Williams 2008). Making this distinction allows one to separately test for different genetic models. More than one genetic model is needed to account for multiple phenotypic patterns of selfed embryo death.

9.5 Deleterious Alleles for Embryo Viability Loci Versus an Embryo Lethal System

For clarity, embryo viability loci (or EVL) is the first model. Here, the process of random mutational accumulation would be expected to generate vast numbers of loss-of-function mutants at embryo viability loci and these mutants are generally partially or fully recessive. Embryo death is caused by recessive homozygotes and rises as this genotypic class increases in direct proportion to inbreeding. Also, death to selfed embryos is expected to occur randomly across all stages.

Heuristic methods have been used to estimate the numbers of embryo viability loci. Such methods are crude yet consistent with other organisms. The total number of EVL range from 75 to 2,000 (Koski 1971; Williams and Savolainen 1996) and this range parallels EVL numbers reported for model organisms (Miklos and Rubin 1996). A higher estimate of 10,000 has also been reported for the Pinaceae (Namkoong and Bishir 1987; Bishir and Namkoong 1987) but one reason for this upper estimate is that these authors assumed an average of 10 lethals, a value which is nearly two times higher than the true average (Kormutak and Lindgren 1996).

The second model is the embryo lethal system (or ELS), shown in Fig. 9.2, which refers to both embryo viability loci and stage-specific or concerted selfed embryo death (Williams 2007). Stage-specific death to selfed embryos was an observation which prompted Koski (1971) to first use the term embryo lethal system. The original studies based on *Picea abies* and *Pinus sylvestris* (Koski 1971) showed a dual death pattern for selfed embryos for several individuals sampled randomly within both species. Some random deaths occurred across all stages but over half of all selfed embryos died at a single stage: early embryogeny.

This same death peak occurred in *Picea abies* which has no cleavage embryos as well as *Pinus sylvestris* which has both simple and cleavage polyembryony (Koski 1971). It coincided with the stage where the dominant embryo elongates into the corrosion cavity (Koski 1971). This death peak is described in greater detail for a recent *Pinus taeda* study (Williams 2008). By defining by viewing models for ELS and EVL separately, one can separate the dual death patterns into two parts and then hypothesize separate genetic models for each component.

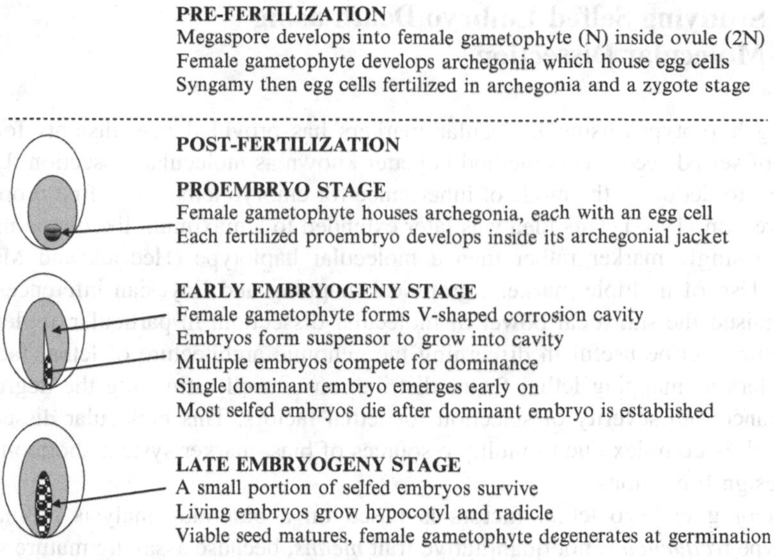

PRE-FERTILIZATION
Megaspore develops into female gametophyte (N) inside ovule (2N)
Female gametophyte develops archegonia which house egg cells
Syngamy then egg cells fertilized in archegonia and a zygote stage
POST-FERTILIZATION
PROEMBRYO STAGE
Female gametophyte houses archegonia, each with an egg cell
Each fertilized proembryo develops inside its archegonial jacket
EARLY EMBRYOGENY STAGE
Female gametophyte forms V-shaped corrosion cavity
Embryos form suspensor to grow into cavity
Multiple embryos compete for dominance
Single dominant embryo emerges early on
Most selfed embryos die after dominant embryo is established
LATE EMBRYOGENY STAGE
A small portion of selfed embryos survive
Living embryos grow hypocotyl and radicle
Viable seed matures, female gametophyte degenerates at germination

Fig. 9.2 Schematic diagram of the embryo lethal system as described by Koski (1971) for *Picea abies* and *Pinus sylvestris*. Selfed embryos die throughout embryo development but death peaks during the early embryogeny stage. Adapted from Williams C. 2008. Selfed embryo death in *Pinus taeda*: a phenotypic profile. New Phytologist 178: 210–222. Copyright permission granted

9.6 Exploring Phylogenetic Limits to the Embryo Lethal System

The phylogenetic limits of the embryo lethal system have yet to be explored. The primary reason is that all organisms accrue deleterious mutations to early-acting viability loci and this accumulation implies a process which is not taxon-specific. But if the EVL model is considered separately from the ELS model then the question of phylogenetic limits becomes relevant for the ELS model. The Pinaceae family is the largest of the seven and all of its taxa are monoecious. To date, the effects of selfing on embryo viability have been studied in only five of the 11 genera within the Pinaceae: *Abies*, *Larix*, *Picea*, *Pinus*, and *Pseudotsuga* (Table 9.2). Selfing effects for the other Pinaceae genera have not yet been reported. Defining phylogenetic limits might not be possible given that some species such as *Pinus resinosa* are self-fertile (Table 9.2).

The question depends on the Cupressaceae family which has received the least study. It is not clear if any additional monoecious genera have extreme inbreeding depression during seed development. As seen in Table 9.2, the only two taxa in this family so far have lethal equivalent estimates and both are self-fertile.

9.7 Studying Selfed Embryo Death using Molecular Dissection

Tracing haplotypes using molecular markers has provided new insights for the death of selfed seeds. This method is better known as molecular dissection. Using markers to deducing the mode of inheritance for embryo lethals was first proposed by Sorensen (1967). This idea was later extended to a maximum likelihood model using a single marker rather than a molecular haplotype (Hedrick and Muona 1990). Use of multiple markers, graphical mapping and Bayesian inferences has since raised the statistical power of molecular dissection. In particular, molecular dissection can be useful in discerning the genomic architecture of lethal factors; this refers to mapping lethal factor distribution as well estimating the degree of dominance and severity of selection for lethal factors. This molecular dissection approach is complex due to multiple sources of bias, marker system shortcomings and design limitations.

Mapping embryo lethal factors is based on a binomial analysis of marker genotype *frequencies*, not quantitative trait *means*, because assaying mature seeds means that DNA can be sampled only for survivors. Severity, degree of dominance and localization of lethal loci is based on a binomial analysis of marker genotype frequencies. Marker alleles tightly linked to lethal factors exhibit transmission ratio distortion (TRD) also known as segregation distortion. TRD is defined as a statistically significant departure from expected Mendelian inheritance regardless of the genetic mechanism (Crow 1991). Testing ratios among marker genotypic classes leads to certain patterns of TRD which can be used to infer the presence of hidden lethal factors co-segregating with one or more markers. If one or more marker genotype classes are under- or over-represented relative to the expected Mendelian segregation ratios of 1:2:1 and other sources of bias can be removed then the degree of dominance and its severity can be estimated for a putative zygotic lethal factor.

Molecular dissection indirectly detects the presence of one or more lethal factors only at the level of large chromosomal segments (Ritland 1996), not at the level of individual genes. Genetic maps partition chromosomes into segments which allow detection of segments correlating with patterns of phenotypic variation or embryo death. Mapping those segments co-segregating with lethal factors yields the linkage map profile of factors causing selfed embryo death, i.e. a crude genomic architecture. Even so, some caveats should be presented first before results from the few experimental studies are discussed.

Molecular dissection is only as valid as its assumptions. Of these, the most critical is the assumption of one marker per lethal factor. This method works well if lethal factors act independently and are distributed across different chromsomes, rather than clustered on a single chromosome or subject to purely epistatic interaction. Another major assumption is that zygotic selection, not gametophytic or gametic selection, is operative and this requires that any bias due to other allele-distorter mechanisms be ruled out.

9.7.1 Sources of Bias for Molecular Dissection

Transmission ratio distortion can be caused by many genetic mechanisms other than deleterious alleles (Bernasconi et al. 2004): pollen killer alleles, meiotic drive alleles, neocentromeres and gametophytic lethals, all of which can skew Mendelian segregation ratios. Such mechanisms constitute a major source of bias when relying on distorted marker alleles as a means of locating embryo lethal factors. Assaying marker ratios in male and female gametophytes prior to fertilization and in nonself crosses can identify and remove bias.

9.7.2 Bias from Gametic or Gametophytic Selection

Seed plants have a gametophyte phase which is absent in animals so gametophytic and gametic selection are used interchangeably in the literature. Gametic selection, or pre-fertilization selection, can mimic various types of zygotic selection, thus biasing dominance estimates and lethal severity if present (Husband and Schemske 1996; Fu and Ritland 1994a,b; Vogl and Xu 2000).

Directional selection, meiotic drive and opposing selection prior to fertilization can all cause an allele to be selectively eliminated or undertransmitted via one or both gametophytes (see review in Williams 2007). This means that gametic ratios can be distorted *prior* to fertilization. One way to detect this form of distortion is to compare the ratios of the two classes of marker heterozygotes A_1A_2 versus A_2A_1 (e.g. Kuang et al. 1999; Williams et al. 2001). These two heterozygote classes can be distinguished by haplotyping female gametophyte tissue and the embryo. The maternal allelic contribution can be deduced as A_1 or A_2. Removing any bias caused by gametic selection is essential before correctly locating zygotic lethal factors.

9.7.3 Marker Systems and Map Density Impose Experimental Design Limitations

Consider the case of balanced lethal systems (Fig. 9.3). Using too few molecular markers for tracing haplotypes, then one is likely to detect apparent overdominance or pseudo-overdominance (Fig. 9.3). Here, the true locus order model shows L_1 and L_2 as two lethal factors linked in repulsion. Each lethal factor is truly dominant, not overdominant, if analyzed correctly using one marker per lethal factor. Otherwise, an excess of heterozygotes (pseudo-overdominance) will be detected even though partial dominance is truly present. It should be noted that clustering for any map, whether saturated or sparse, can bias detection, estimation and analyses. Detecting clustered lethals requires physical or cytogenetic mapping.

More highly saturated maps are available using dominant markers. These band-present, band-absent systems have higher polymorphism levels and thus provide

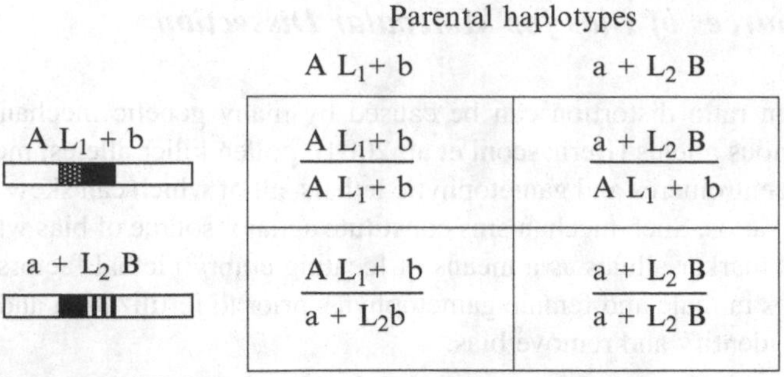

Fig. 9.3 A balanced lethal system, also known as pseudo-overdominance, refers to two lethal factors linked in repulsion-phase as show here. Assuming the simplified case of close linkage (c=0), then gametes have one of two parental haplotypes in the case of selfing (i.e. no recombinants). With this balanced lethal system, only the double marker heterozygote is recovered. Other marker genotypic classes are not recovered because they are homozygous for one of the two lethals. Each lethal factor is dominant rather than overdominant but the tight linkage causes apparent overdominance to be detected instead. Williams, C. (2007) Re-thinking the embryo lethal system within the Pinaceae. Canadian Journal of Botany 85: 667–677. Copyright permission granted

higher map density. The tradeoff is that this marker type provides no direct estimates of heterozygosity although some of the lost information for dominant markers can be offset by genotyping each embryo's female gametophyte (Kuang et al. 1999). Ideally, better hypervariable, codominant marker systems could mean better resolution of the genomic architecture for zygotic lethal factors.

Another source of bias comes with using single markers for the detection of zygotic lethal factors. Here, the magnitude of the lethal factors is confounded with the linkage distance between the marker and the lethal factor. The more distant the marker, then the smaller the lethal factor effect becomes. Single-marker detection has another drawback: it requires *a priori* assumptions about the degree of dominance. With only two degrees of freedom, degree of dominance cannot be estimated so the simplest case, complete dominance ($h = 0$), is often assumed (Hedrick and Muona 1990). With this model, observed frequencies of the three genotypic classes from selfed progeny are used to jointly estimate the selection coefficient and the recombination fraction between the marker and its hidden or putative zygotic lethal factor.

An alternative to the single-marker problem is to use the model-free graphical representation method (Fu and Ritland 1994a,b). This can be used with a single-marker approach. Genotypic ratios of markers linked to a putative lethal factor are placed on the graph and this provides the estimate of dominance. This method simply graphically partitions the selection space among all possible marker segregation ratios and provides inferences about putative lethal factors (Fu and Ritland 1994a,b).

This method also has its own drawbacks and these include 1) selection space overlaps for different degrees of dominance, yielding some ambiguity, 2) the method provides no error estimates (Kärkkäinen et al. 1999) and 3) both recombination fractions and selection coefficients between a marker and its putative lethal locus must be assumed rather than estimated (Fu and Ritland 1994a,b). Here too, gametophytic or gametic selection must be detected independently (Vogl and Xu 2000). Nonetheless results from graphical representation have been robust when compared to a Bayesian approach (Kärkkäinen et al. 1999).

9.8 Genomic Architecture for Zygotic Lethal Factors

Embryo death over the course of *Pinus* spp. development is caused by many semi-lethal factors (which is defined as selection coefficient or s = 0.5). In molecular dissection studies for diploxylon *Pinus* spp, factors influencing embryo death are located across the entire genome. Oddly, marker density appears to be curvilinear with respect to the number of zygotic lethal factors (Williams 2007) which suggests distribution across few different chromosomal locations. For example, nine different lethal factors were detected with 150 RAPD markers for *Pinus radiata* (Kuang et al. 1999) and this value double to 20 lethal factors when the number of markers was increased to 226 AFLP markers in *Pinus taeda* (Remington and O'Malley 2000a). One lethal factor was detected for every two markers in a third study using codominant microsatellites (Williams et al. 2003) although this was a sparse map and prone to detecting more than one zygotic lethal per chromosomal segment. The likely explanation is that some zygotic lethal factors are located in balanced lethal complexes (Fig. 9.3).

To date, semi-lethal factors influencing embryo death have been reported across studies on *Pinus radiata* and *Pinus taeda*, respectively (Kuang et al. 1999; Remington and O'Malley 2000a,b; Williams et al. 2001) using interval mapping. Fully lethal loci (s = 1) are extremely rare in *Pinus radiata* seed and seedlings (Kuang et al. 1999). Fully lethal effects are rare and this finding runs contrary to the meta-analysis predictions based a wide range of angiosperm plants (Husband and Schemske 1996). One interesting finding is that factors influencing embryo death are not pleiotropic or co-expressed during growth in older seedlings. This implies that the same mutational defects are not continually expressed throughout the young sporophyte's early growth trajectory (Remington and O'Malley 2000b).

Most semi-lethal factors for *Pinus radiata* and *Pinus taeda* are also partially dominant (Kuang et al. 1999; Remington and O'Malley 2000a). These results contrasts with Williams et al. (2003) where overdominant factors, detected using codominant markers, were more prevalent. One overdominant zygotic lethal factor has transient expression coinciding with embryo separation from its female gametophyte (Williams et al. 2001). In these studies, overdominant lethal factors are tightly linked to both common and rare marker alleles and gametic selection has been ruled out (Williams et al. 2003). Molecular dissection studies, although cannot rule out the balanced lethal systems shown in Fig. 9.3, show heterogeneity among factors with respect to lethality, dominance and timing of expression.

Perhaps molecular dissection is limited to crude approximation of the number of unlinked segments. It is not likely to be informative about clustered lethals, specific genes or family of genes or even a regulatory element or co-adapted gene complexes within a given chromosomal segment. Molecular dissection is a starting place.

9.9 Closing

To solve the riddle of Plate III, all of the preceding chapters must be applied. Why is this considered to be an anomaly? Two reasons. First, it is surprising that embryo competition was so mild that two embryos survived, selfed or not. Second, the odds are slim that any *Pinus elliottii* selfed embryo will survive past seed maturity given high inbreeding depression for this species during seed development. And so this hypothetical model was explained using molecular markers and haplotypes, providing an introduction to the more complex concept of molecular dissection.

Proposing – and testing - other models for zygotic lethals or even testing new genetic models to account for the embryo lethal system is now possible by combining experimental tools such as histology, magnetic resonance imaging, genomics and proteomics. No doubt that random mutational accumulation accounts for some selfed embryo death but additional explanations should be sought for the death peak, experimental and theoretical alike. Is the dual death pattern for selfed embryo observed for other conifer taxa or not?

Understanding the genetic models behind selfed embryo death has value beyond conifer reproductive biology. Understanding the genetic basis for selfed embryo death in the Pinaceae is germane to a broad range of eukaryotic organisms, not just Pinaceae, conifers or plants. Knowing its genetic basis, or even refining action attributed to embryo viability loci, can refine theoretical models about the genetic basis of inbreeding depression. Better models in turn provide insights for better management of closed or captive populations.

References

Bernasconi G., T. Ashman et al. 2004. Evolutionary ecology of the prezygotic stage. Science 303: 971–975.

Bishir, J. and G. Namkoong. 1987. Unsound seed in conifers: estimation of numbers of lethal alleles and of magnitudes of effects associated with the maternal parent. Silvae Genetica 36: 180–185.

Bramlett, D. and F. Bridgwater. 1986. Segregation of recessive embryonic lethal alleles in a F1 population of Virginia pine. IUFRO Conference on Breeding Theory, Williamburg VA.

Bramlett, D. and T. Popham. 1971. Model relating unsound seed and embryonic lethals in self-pollinated pines. Silvae Genetica 20: 192–193.

Buchholz J. 1918. Suspensor and early embryo of *Pinus*. Botantical Gazette 66: 185–228.

Buchholz, J. 1920. Embryo development and polyembryony in relation to the phylogeny of conifers. American Journal of Botany 7: 125–145.

Buchholz, J. 1922. Developmental selection in vascular plants. Bot. Gaz. 73: 249–286.

Cavalli-Sforza, L. and W. Bodmer. 1971. *The genetics of human populations*. San Francisco, Freeman.

Crow, J. 1991. Why is Mendelian segregation so exact? Bioessays 13: 305–312.

Dogra, P. 1967. Seed sterility and disturbances in embryogeny in conifers with particular reference to seed testing and tree breeding in Pinaceae. Studia Forestalia Suecica 45: 1–97.

Filonova L.H., von Arnold S. et al. 2002. Programmed cell death eliminates all but one embryo in a polyembryonic plant seed. Cell Death and Differentiation 9: 1057–1062.

Fowler, D. 1965a. Natural self-fertilization in three jack pines and its implications in seed orchard management. For. Sci. 11: 55–58.

Fowler, D. 1965b. Effects of inbreeding in red pine, *Pinus resinosa* Ait. Silv. Genet. 12: 12–23.

Franklin, E. 1969. Inbreeding depression in metrical traits of loblolly pine (*Pinus taeda* L.) as a result of self-pollination. Ph.D. Dissertation, Raleigh NC., School of Forest Resources, North Carolina State University.

Franklin, E. 1972. Genetic load in loblolly pine. Amer. Nat. 106: 262–265.

Fu, Y-B and K. Ritland. 1994a. Evidence for the partial dominance of viability of viability genes contributing to inbreeding depression in *Mimulus guttatus*. Genet. 136: 323–331.

Fu, Y-B and K. Ritland 1994b. On estimating the linkage of marker genes to viability genes controlling inbreeding depression. Theor. Appl. Genet. 88: 925–932.

Fu, Y.B., D. Charlesworth et al. 1997. Point estimation and graphical inference of marginal dominance for two viability loci controlling inbreeding depression. Genet. Res. 70: 143–153.

Gifford, E. and A. Foster. 1989. *Morphology and evolution of vascular plants*. W.H. Freeman Company, New York.

Griffin, R. and D. Lindgren. 1985. Effect of inbreeding on production of filled seed in *Pinus radiata* - experimental results and a model of gene action. Theor. Appl. Genet. 71: 334–343.

Hagman, M. and L. Mikkola. 1963. Observations on cross-, self- and interspecific pollinations in *Pinus peuce* Griseb. Silv. Genet. 12: 73–79.

Haig D. 1992. Brood reduction in gymnosperms. In *Cannibalism: Ecology and Evolution among Diverse Taxa*. Edited by M. Elgar and B. Crespi. pp. 63–84. Oxford University Press.

Hauser, T. and V. Loeschcke. 1994. Inbreeding depression and mating distance dependent offspring fitness in large and small populations of *Lynchis floscuculi* (Caryophyllaceae). J. Evol. Biol. 7: 609–622.

Hedrick, P. and O. Muona. 1990. Linkage of viability genes to marker loci in selfing organisms. Hered. 64: 67–72.

Husband, B. and D. Schemske. 1996. Evolution and timing of inbreeding depression in plants. Evol. 50: 54–70.

Kärkkäinen, K., H. Kuittinen et al. 1999. Genetic basis of inbreeding depression in *Arabis petraea*. Evol. 53: 1354–1365.

Kang, H., C. Hardner et al. 1994. Lethal loci and lethal equivalents in willow, *Salix viminalis*. Silv. Genet. 43: 138–145.

Kormutak, A. and D. Lindgren. 1996. Mating system and empty seed in silver fir (*Abies alba* Mill.). For. Genet. 3: 231–235.

Koski, V. 1971. Embryonic lethals of *Picea abies* and *Pinus sylvestris*. Commun. Institute of Forestalia Fennica 75: 1–30.

Koski, V. 1973. On self-pollination, genetic load and subsequent inbreeding in some conifers. Communicationes Instituti Forestalis Fenniae 78: 1–42.

Krebs, S. and J. Hancock. 1991. Embryonic genetic load in the highbush blueberry, *Vaccinium corymbosum* (Ericaceae). Amer. J. Bot. 78: 1427–1437.

Kuang, H., T. Richardson et al. 1999. Genetic analysis of inbreeding depression in plus tree 850.55 of *Pinus radiata* D. Don. II. Genetics of viability genes. Theor. Appl. Genet. 99: 140–146.

Lande, R. and D. Schemske. 1985. The evolution of self-fertilization and inbreeding depression in plants. I. Genetic models. Evol. 39: 24–40.

Lande, R., D. Schemske et al. 1994. High inbreeding depression, selective interference among loci, and the threshold selfing rate for purging recessive lethal mutations. Evol. 48: 965–978.

Latta, R. and K. Ritland. 1994. The relationship between inbreeding depression and prior inbreeding among populations of four *Mimulus* taxa. Evol. 48: 806–817.

Levin, D. 1991. The effect of inbreeding on seed survivorship in *Phlox*. Evol. 45: 1047–1049.

Lewontin, R. 1974. *The genetic basis of evolutionary change*. NY, Columbia University Press.

Libby, W., B. McCutchan et al. 1981. Inbreeding depression in selfs of redwood. Silv. Genet. 30: 15–25.

Lindgren, D. 1975. The relationship between self-fertilization, empty seeds and seeds originating from selfing as a consequence of polyembryony. Studia Forestalia Suecica 126: 1–24.

McCune, A., R. Fuller et al. 2002. A low genomic number of recessive lethals in natural populations of bluefin killifish and zebrafish. Science 296: 2398–2401.

Mergen F., J. Burley et al.1965. Embryo and seedling development in *Picea glauca* (Moench) Voss after self-, cross-, and wind-pollination. Silv. Genet. 14: 188–194.

Miklos G. and G. Rubin. 1996. The role of genome project in determining gene function insights from model organisms. Cell 86: 521–529.

Morton, N., J. Crow et al.1956. An estimate of the mutational damage in man from data on consanguineous marriages. Proc. Natl Acad. USA 42: 855–863.

Muller, H. 1950. Our load of mutations. Am. J. Hum. Genetics 2: 111–176.

Namkoong, G. and J. Bishir. 1987. Frequency of lethal alleles in forest tree populations. Evol. 41: 1123–1127.

O'Connell, L. and Ritland K. 2005. Post-pollination mechanisms promoting outcrossing in a self-fertile conifer, *Thuja plicata* (Cupressaceae). Can.J. Bot. 83: 335–342.

Orr-Ewing, A. 1957. A cytological study of the effects of self-pollination on *Pseudotsuga menziesii* (Mirb.) Franco. Silv. Genet. 6: 179–185.

Ouborg, N. and R. von Treuren. 1994. The significance of genetic erosion in the process of extinction. IV. Inbreeding load and heterosis in relation to population size in the mint *Salvia pratensis*. Evol. 48: 996–1008.

Owens J., A. Colangeli et al. 1990. The effect of self-, cross- and no pollination in ovule, embryo, seed and cone development in westen red cedar (*Thuja plicata*). Can.J. For. Res. 20: 66–75.

Park, Y. and D. Fowler. 1982. Effects of inbreeding and genetic variances in a natural population of tamarack (*Larix laricina* (Du Roi) K. Koch) in eastern Canada. Silv. Genet. 31: 21–26.

Porcher E. and Lande R. 2005. Reproductive compensation in the evolution of plant mating systems. New Phytol. 166: 673–684.

Ralls, K., J. Ballou et al. 1988. Estimates of lethal equivalents and the cost of inbreeding in mammals. Conserv. Biol. 2: 185–193.

Remington D. and D. O'Malley. 2000a. Whole-genome characterization of embryonic stage inbreeding depression in a selfed loblolly pine family. Genet. 155: 337–348.

Remington, D. and D. O'Malley 2000b. Evaluation of major genetic loci contributing to inbreeding depression for survival and early growth in a selfed family of *Pinus taeda*. Evol. 54: 1580–1589.

Ritland K. 1996. Inferring the genetic basis of inbreeding depression in plants. Genome 39: 1–8.

Sarvas, R. 1962. Investigations on the flowering and seed crop of *Pinus silvestris*. Institute Forestalis Fennica Comm. 53: 1–198.

Savolainen, O., K. Kärkkäinen et al. 1992. Estimating numbers of embryonic lethals in conifers. Heredity 69: 308–314.

Sittman, K., B. Abplanalp et al. 1966. Inbreeding depression in the Japanese quail. Genetics 54: 371–379.

Skinner D. 1992. Ovule and embryo development, seed production and germination in orchard grown control pollinated loblolly pine (*Pinus taeda* L.) from coastal South Carolina. M.Sc. Thesis, University of Victoria, Victoria B.C. Canada.

Sorensen, F. 1967. Linkage between marker genes and embryonic lethal factors may cause disturbed segregation ratios. Silv. Genet. 16: 132–134.

Sorensen, F. 1969. Embryonic genetic load in coastal Douglas fir, *Pseudotsuga menziesii* var. *menziesii*. Amer. Nat. 103: 389–398.

Sorensen, F. 1971. Estimate of self-fertility of Douglas-fir from inbreeding studies. Silv. Genet. 20: 115–120.

Sorensen, F. 1982. The roles of polyembryony and embryo viability in the genetic system of conifers. Evol. 36: 725–733.

Vogl C. and S. Xu. 2000. Multiple-point mapping of viability and segregation distorting loci using molecular markers. Genetics 155: 1439–1447.

Williams, C., R. Barnes et al. 1999. Embryonic lethal load for a neotropical conifer, *Pinus patula* Schiede and Deppe. J. Hered. 90: 394–398.

Williams, C. and O. Savolainen. 1996. Inbreeding depression in conifers: implications for breeding strategy. For. Sci. 42: 102–117.

Williams, C., Y. Zhou et al. 2001. A chromosomal region promoting outcrossing in a conifer. Genetics 159: 1283–1289.

Williams, C., L. Auckland et al. 2003. Overdominant lethals as part of the conifer embryo lethal system. Heredity 91: 584–592.

Williams, C. 2007. Re-thinking the embryo lethal system within the Pinaceae. Canadian Journal of Botany 85: 667–677.

Williams C. 2008. Selfed embryo death in *Pinus taeda*: a phenotypic profile. New Phytologist 178: 210–222.

Conclusion

Humans have long held a practical interest in conifers but few realize the lengthy evolutionary history of a conifer or how peculiar its reproduction truly is. But it is here that one can find a wealth of interesting research questions.

The first question concerns the role for reproductive biology in the persistence of conifers. These plants have persisted through radical climate change, dinosaur grazing and repeated ecological upheaval only to be extirpated (or for few, displaced) by another group of seed plants, the angiosperms. Indeed, the discovery of Wollemi pine should be celebrated as a global treasure and it seems appropriate that schoolchildren in Australia plant its seedlings in botanical gardens. This conifer lives even though dinosaurs died and the earth's first Paleozoic forests are long gone. Might the unusual yet abundant reproduction in conifers have contributed to such persistence?

The second question concerns the adaptive importance of the diplohaplontic life cycle itself. This book about conifer reproduction is framed around the diplohaplontic life cycle. Here one can see that heterospory and the seed habit are profound features but it is not well understood that the gametophyte is an independent multicellular organism not a type of gamete. To add further to this confusion, consider the case of paternal apomixis where Algerian cypress simply skips the haploid stage altogether, forming no male gametes at all. Conifer reproductive biology is peculiar both as rule and as exception but even so, it is not clear how both diploid and haploid phases both persist.

The third question is concerned with why conifers only have a single mating system, wind-pollination, yet flowering plants have many mating systems along with a variety of pollination vectors other than wind. Does this conifer singularity have a genomic explanation? My working hypothesis is that conifers have genomic mechanism(s) which suppress gross chromosomal changes, i.e. karyotypic orthoselection. If so, then such genomic conservatism would begin with the recombination modification system. The longheld view is that conifer karyotypes evolve more slowly than other plants but the more compelling view is that conifer genomes have mechanisms that actively select against radical change in reproduction.

The fourth question concerns the observation that each female gametophyte has multiple egg cells. This observation has sustained interest across three centuries of Western scientific thought. But it is now widely understood that the female gametophyte and its multiple archegonia is a conserved feature for both extant and extinct seed plants. Even the earliest seed plant fossils, the hydrasperman plants, had multiple archegonia. Angiosperms are thus the exception, not the rule.

The fifth question concerns the buoyancy of the male gametophyte. The male gametophyte's capacity for long-distance windborne transport has been the subject of scientific discourse for over 100 years. But how far? The world's record is 1,000 km from source, as Gunnar Erdtman showed with his transatlantic vacuum cleaners in 1937. And a few taxa have pollen that can even float, thus endowed with the aerodynamics of a floatplane. And conifer pollen is hard to kill; Douglas fir pollen grain can still germinate after being frozen for years. There can be no doubt about the high nuisance value of conifer pollen. This is particularly true for those few conifer taxa which trigger severe respiratory allergies in humans.

The sixth question concerns the mystery of the pollination drop. As a result, female and male gametophytes approach slowly rather than collide. The *Juniperus communis* pollination drop ceases to emerge a second time after it captures its own pollen. But the drop does not respond in the same way to pollen of other seed plants nor silica particles of the same size. And the pollination drop expresses fungal resistance proteins, protecting pollen, the nucellus and perhaps the future female gametophyte from pathogen attack. From motile sperm delivery to pollination, this aqueous fluid that emerges under the cover of night prompts interesting research questions about its origin and its molecular cues.

The seventh question concerns the surprising degree of divergence between the conifer seed and the angiosperm seed. Even so, both types of seeds are prone to hitchhiking as a means of long-distance transport and they contain more than one phase of the life cycle. The edible pine seed has been dispersed by birds, Neanderthals and Phoenicians. From pollen tube delivery, syngamy, organelle inheritance and subsequent zygote to embryo development, these stages in particular show to how distinct conifers are from angiosperms. The edible pine seed, composed of the haploid female gametophyte and the diploid embryo, is a reminder of the diplohaplontic life cycle itself.

And the final question concerns how the working parts of the mating system fit together. Conifers, although sessile plants, do exert mate choice and they are usually outcrossing although exceptions exist. Selfed pollinations might be avoided by temporal and spatial separations of male and female strobili in the crown of a monoecious species but this is an incomplete barrier and such accidents are eliminated after fertilization. Outcrossing with a sympatric relative can produce pollen arrest but more often produces a seedling which grows into a fertile F1 adult.

One might consider these three research topics in greater depth:

Research Topic 1: Plant biology textbooks present the seed plant life cycle as a steady reduction of the haploid gametophyte phase coupled with an increasing dominance of the diploid sporophyte phase. While this is largely true, it obscures deeper questions yet to be answered in life cycle evolution. Theoretical advances are providing a steady stream of elegant models but experimental work, especially for conifers, lags far behind. This suggestion runs parallel to Mable and Otto's (1998) statement that "life cycle strategy can be viewed as a selectively variable trait capable of continued evolutionary modification." This avenue of research for conifers can be posited as three specific questions:

- What is the adaptive benefit for maintaining alternating phases where both haploid and diploid phases have complex, synchronous roles? Mable and Otto (1998) outline a number of testable hypotheses but the masking hypothesis is the simplest explanation for Occam's razor: *diploid tissues mask deleterious mutations while haploid tissues act as a barrier*. To this, I would add that for the conifers, the haploid tissues act as a partial barrier or perhaps a selective sieve for *selecting only against those mildly deleterious mutations affecting basic cell functions and cell-to-cell competition*. If more specific, one can further hypo-thesize that it is the female gametophyte, with its larger numbers of cell divisions and greater longevity, that might be a more effective selective sieve than the male gametophyte. This is an argument for male-driven mutation; some interesting support has been marshaled using organellar DNA in gymnosperms (Whittle and Johnson 2003). And one should note that the search for somatic mutations has begun with female rather than male reproduction (Cloutier et al. 2003).
- Is karyotypic orthoselection operative for conifer genomes? And if so, what are mechanisms? Do they influence the stability of the diploid–haploid life cycle for conifers? One thought is that such stringent genome-conserving mechanisms might impede adaptive radiation even when presented with new selection space. Contrast wind-pollination, the sole mating system for conifers, with the myriad of mating systems – and animal pollen vectors – found among angiosperms. In angiosperms, such mating system variability coincides with a tolerance of gross chromosomal rearrangements and a general absence of karyotypic orthoselection, especially at the taxonomic family level.
- What are the molecular cues and recognition systems between the sporophyte and its gametophyte phases? Two examples come to mind. Consider what cues occur during pollination: a synchronous, highly orchestrated set of actions and reactions take place between sporophytic and gametophytic tissues. Scales on a female strobilus flex open, a pollination drop emerges from an ovule at the scale's base, captures pollen then does not emerge again. Similarly, fertilization events also rely on cues between phases; the male gametophyte's pollen tube resumes its path through the sporophyte's nucellus only days before fertilization.

Research Topic 2: The author is influenced by twentieth century paleobotanists and other "students of death" (Niklas 1997) who compare vascular plant reproductive form and function from the Silurian to the Mesozoic era. The advent of separate female and male spores opened the way to divergence in sporangial cell lineages. For female sporogenesis, only one megaspore survived to develop into a colorless, endosporic gametophyte. The sporophyte's integument and its micropyle have also undergone radical changes; heterospory has had consequences for strobilus formation well before sporogenesis (Niklas 1997) and for the evolution of the seed habit. By contrast, male sporogenesis leads to the survival of all four microspore yet each microspore divides into a few cells via asymmetric mitoses. Upon capture, airborne pollen grains germinate a specialized distal pollen tube for the two-sperm delivery.

Heterospory's divergence is bounded by a requirement for precision events leading up to pollination. Synchrony between male and female parts is highly coordinated towards ensuring pollination success. Optimizing pollen release while ensuring pollen capture is the shared target for male and female reproduction whether the plant is monoecious or dioecious. This opens questions about what type of regulatory cues or recognition systems must occur between female and male reproductive structures at pollination – and fertilization. Evolution of conifer reproduction can be viewed as parabolic: while heterospory opened selective space for sex-specific divergence in form and function, the reliance on a wind-pollination system brings precise requirements for synchronous male and female developments events as a hedge against the uncertainty of aerial transport.

Making mate choices is another example of finding an optimal balance between divergence and synchrony. Pollen choice cannot be too closely related or too distantly related Avoiding selfing in favor of outcrossed pollen can result in capturing a source of interspecific pollen. Here too, crosstalk between female and male gametophytes must somehow occur in order to reduce these odds of failing to secure a mate choice. What are the means by which conifers exert mate choice? What causes pollen tube arrest? What causes death to selfed embryos? The communications network among all of the working parts within this dynamic, peculiar conifer mating system need full molecular characterization.

Research Topic 3: *Bauplan* hypotheses. Remember that conifers are not grouped according to a shared set of reproductive characters but rather by the absence of characters. Without a more representative knowledge of conifer reproductive biology as a whole (i.e. Southern Hemisphere conifers, those less accessible taxa worldwide in all families and the newly discovered species *Xanthocyparis vietnamensis*) then this first iteration must be viewed as a set of definitive testable hypotheses.

Female gametophyte: Few features are truly conserved. Those more highly conserved than others are the stages of development starting from the free nuclear stage (with the *Sequoia* exception) through the formation of multiple archegonia (with the *Taxus* exception). Even so, monospory and the total numbers of female gametophyte cells are not conserved. Perhaps the most interesting of the conserved features is the long interval between pollination to fertilization. Here too, the absolute duration varies among taxa but all conifer taxa studied to date have this lengthy interval during which the female gametophyte completes its development.

Male gametophyte: The polarity of the tetrad, its asymmetric mitoses and the presence of a pollen wall, the distal aperture, the siphonogamous pollen tube, its hydration requirement and diplospermy are all shared characters. But these shared features are conserved among all modern seed plants, not only conifers (Rudall and Bateman 2007). An interesting feature for further research is the RNA and protein expression patterns from microspore formation to male gametophyte cell division. Some members of the Pinaceae lack protein expression in the interval between pollen release and germination (Pettit 1985). Does this hold for all conifer taxa, including the more exceptional members of the Southern Hemisphere conifer families? What about transcript partitioning among cell types within the male gametophyte? Transcript partitioning is the case for angiosperm species such as *Zea mays* (Engel et al. 2003). As a general rule, male reproduction tolerates a high degree of variability whether it is pollen wall formation, response to hydration, total number of mitotic divisions and how (or where) the pollen germinates. Comparative female–male studies on the same species are also rarely reported.

Sporophytes: Embryo development of the young sporophyte progresses through three stages in the most conifer taxa. The reproductive structures and tissues on the adult sporophyte, female and male, show a high degree of variability but a few exceptions deserve mention. The micropyle-chalazal polarity of the ovule, the megaspore wall around the female gametophyte and the location for sporogenesis itself are highly conserved characters. The *Bauplan* concept, whether conifer-specific or not, is useful for elucidating the communications network within a diplohaplontic life cycle persisting over the course of 300 million years.

In conclusion, it seems trivial but necessary to point out that this book and its suggested research ideas have been made possible not only by this author but also from the generosity of life science scholars for hundreds of years. We all stand on the shoulders of titans.

References

Cloutier, D., D. Rioux, et al. 2003. Somatic stability of microsatellite loci in eastern white pine, *Pinus strobus* L. Heredity 90: 247–252.

Engel, M., A. Chaboud, et al. 2003. Sperm cells of *Zea mays* may have a complex complement of mRNAs. Plant Journal 34: 697–707.

Mable, B. and S. Otto. 1998. The evolution of life cycles with haploid and diploid phases. BioEssays 20: 453–462.

Niklas, K. 1997. *The Evolutionary Biology of Plants*. University of Chicago Press, Chicago, IL.

Pettit, J. 1985. Pollen tube development and characteristics of the protein emission in conifers. Annals of Botany 56: 379–397.

Ruddall, P. and R. Bateman. 2007. Developmental bases for key innovations in the seed-plant microgametophyte. Trends in Plant Science 12: 317–326.

Whittle, C.-A. and M. Johnston. 2003. Male-driven evolution of mitochondrial and chloroplastidial DNA sequences in plants. Molecular Biology and Evolution 19: 938–949.

Glossary

alveolus, alveoli (pl.): refers to a honeycomb-like cavity that forms at the end of the free nuclear stage of female gametophyte development.

antheridial cell: derivative of the antheridial initial during male gametophyte development; also defined as generative cell or spermatogenous cell. The antheridial cell divides into a body cell and a stalk cell.

antheridial initial: derivative of the central cell during development of the male gametophyte; the antheridial initial divides into the antheridial cell and the tube cell.

apical meristem: undifferentiated vegetative branch tip.

apomixis: embryo develops occurs without fertilization.

apoptosis: programmed cell death.

archegonium, archegonia (pl.): the multicellular covering around each egg cell within the female gametophyte.

body cell: fertile derivative of the antheridial cell during development of the male gametophyte. The body cell divides into two male gametes (diplospermy).

bryophytes: spore-bearing plants with independent haploid gametophytic and diploid sporophytic phases where the gametophyte is dominant.

C-value: genome size for an individual or taxon, measured in picograms.

central cell: immediate derivative of the microspore in development of the male gametophyte; central cell divides into the antheridial initial and a prothallial cell.

chalaza, chalazal: juncture where nucellus and integument join, near the stalk; the pole opposite of the micropyle.

chiasma, chiasmata (pl.): a process inherent to meiotic recombination where two homologous chromatids exchange reciprocal DNA, i.e. a crossover.

cone: a female strobilus after pollination and after fertilization.

conelet: a female strobilus after pollination and prior to fertilization.

corrosion cavity: refers to the center of the female gametophyte which breaks down during embryo development to form an opening.

cupule: a cup-like structure that partially covered the preovule on hydrasperman plants.

dichogamy: asynchronous male and female strobilus development which minimized probability of selfing.

dioecy: separate male and female strobili on different plants.

diplospermy: production of two male gametes, both of which are derived from the same generative or spermatogenous cell.

distal face: one polar end of a male microspore. At the end of male meiosis, a rounded tetrad of four microspores forms; each microspore has an outward-facing (distal) pole. Its polarity shapes subsequent development into a pollen grain. See proximal.

embryo lethal system: self-pollinated embryos die during embryo development; this enigmatic phenomenon has been collectively defined as the embryo lethal system.

endospory: enclosure of the mature gametophyte within a spore wall. Usually refers to the female gametophyte inside the ovule.

exine: outermost layer of pollen wall, mostly composed of sporopollenin.

geitonogamy: self-pollination between separate male and female reproductive structures on the same plant (not autogamy).

haploidy: one-half of the chromosomal complement.

haustoria: structures that acquire nutrients from other organisms by tissue penetration.

heterospory: condition of separate male (microspores) and female spores (megaspores).

homogamic hybrid complex: a type of hybridity which results in fertile hybrids which in turn mate with other hybrids or parental species.

homospory: condition of a single spore type for both male and female gametes.

hydrasperman: an extinct form of gymnosperm reproduction where the ovule (or preovule) has a lagenostome, cupules and integumentary lobes but lacks true micropyle.

hypocotyl: the portion of the developing embryo between its radicle and cotyledons.

integument: sheathing or covering around the ovule's nucellus which begins at chalazal end then forms the micropylar opening.

intine: innermost layer of pollen wall, mostly composed of pectin and cellulose.

iteroparity, iteroparous: capable of producing successive cohorts of offspring.

karyotype: the chromosomal complement of the individual.

lagenostome: a funnel-shaped pollen trap above the pollen chamber in hydrasperman plants.

leptoma: the opening or aperture for the germination tube.

meiosis: process by which chromosomal number is halved.

megagametophyte: haploid, multicellular female gametophyte which usually forms from a single megaspore.

megasporangium: integumented ovule.

megaspore mother cell: sporogenous cell which undergoes female meiosis, yielding four megaspores.

megaspore wall: covering for the female gametophyte inside the ovule; derived from diploid tissue of the adult sporophyte.

megasporogenesis: the process by which the megaspore mother cell undergoes female meiosis.

micropyle, micropylar: an opening in the ovule through which pollen enters; an opening formed by the integument covering the ovule.

microsporangia: pollen sacs borne on the abaxial side of a microsporophyll of the male strobilus.

microspore: first cell of the male gametophyte which later becomes multicellular. The microspore divides into the central cell and a prothallial cell.

microsporogenesis: developmental process which produces four haploid microspores from a diploid microsporocyte via meiosis.

microspore mother cell: sporogenous cell which undergoes male meiosis yielding four haploid microspores.

microsporophyll: modified leaves on a male strobilus to which male sporangial sacs attach on the underside.

mitochondria: organelle critical to cell metabolism characterized by multiple copies of a circular DNA genome or haplotype.

monecy, monecious: separate male and female structures both of which occur on the same plant.

monospory, monosporic: in female meiosis, this is the case where only one of the four meiotic products or megaspores survives then develops into a multicellular haploid female gametophyte.

monozygotic: originating from one fertilized embryo; see polyembryony, cleavage.

nexine: the inner layer of the exine around the male gametophyte which acts as an ultrafilter membrane. See sexine.

nucellus: part of the ovule derived from the adult sporophyte; spongy tissue covered by the integument.

ovule: integumented ovule; a structure attached to the female strobilus which includes the integument, the nucellus and the megaspore mother cell (which develops into the female gametophyte and its egg cells). Upon fertilization, the ovule becomes a developing seed.

paleobotany: study of fossil plants.

plastid: a pigment-rich organelle critical to photosynthesis characterized by many copies of a circular DNA genome or haplotype in seed plants.

polarity: the polar alignment of the female ovule and male microspore is set early in development then shapes all subsequent events.

pollen: male gametophyte enclosed in a pollen wall; microspores with a germination tube emerging from the distal face.

pollen wall: this secreted enclosure surrounds the multicellular male gametophyte to form a pollen grain. Secreted by tapetal cells surrounding each pollen mother cell inside the microsporangial sac, the porous wall is composed of sporopollenin and other substances.

pollen tube: arises from the inner layer of the pollen wall or the intine then emerges from aperture or leptoma on distal face of pollen grain before growing between nucellar cells inside the ovule.

pollination drop: aqueous, protein-rich substance secreted at the micropyle of the ovule during pollination which then retracts upon pollen capture.

polyembryony, simple: fertilization of multiple egg cells within a single female gametophyte, results in polyzygotic embryos.

polyembryony, cleavage: fertilization of a single egg cell which later splits into multiple embryos; results in monozygotic embryos.

polyzygotic: originating from multiple zygotes; see polyembryony, simple.

preovule: ovule lacking a true integumentary micropyle typical of extinct hydrasperman plants.

progymnosperms: a group of pteridophytic (fern0like) plants which has secondary cambium and thus the capacity to produce secondary xylem. Generally heterosporous, they often had elaborate branching patterns.

prepollen: microspores which have a proximal opening for sperm release. Typical of early seed plants and Walchian conifers.

proembryo: first major stage of development where the embryo grows within its archegonium.

primary parietal cells: cell layers surrounding the sporogenous cells which give rise to either pollen mother cells or megaspore mother cells; in male sporangia, parietal cells are progenitors of cells which function as the tapetum.

progymnosperms: group of extinct plants with a fern-like or pteridophytic reproduction and gymnosperm secondary wood.

prothallial cells: the sterile cells cut off by the microspore (primary prothallial cell) and again by central cell (secondary prothallial cell) during development of the male gametophyte.

proximal: at the end of male meiosis, each microspore in the tetrad has an inward-facing (proximal) pole. Its polarity shapes subsequent development. See distal.

pteridophytes: fern-like; spore-bearing plants with haploid gametophyte and diploid sporophyte phases; sporophyte is dominant.

pteridosperms: seed ferns. Early pteridosperms were either hydrasperman (lyginopterids) or medullosan. Ovules were radially symmetric.

radicle: primordial root on the developing embryo.

recombination, meiotic: collective term for the processes of DNA exchange, such as unequal (gene conversion) and reciprocal (crossing over).

saccus, sacci (pl.): one or more air-filled wings develop on each microspore within a tetrad; these enlarge on the face of each microspore which lies towards the periphery (or distal end) of the tetrad.

salpinx: in extinct hydrasperm plants which lacked a true micropylar opening, this was a funnel-shaped structure attached to the megasporangium which gave microspore access.

semelparity, semelparous: capable of producing offspring once in a lifetime in perennial plants.

sexine: outer layer of the exine covering the male gametophyte which acts as a porous filter. See nexine.

siphonogamy: delivery of male gametes to the ovule via a germination tube.

spermatophytes: heterosporous plants bearing seeds.

spermatozoid: a motile flagellated male gamete.

sporogenesis: process of reproduction by spores.

sporophyte: the diploid stage in a eukaryotic life cycle.

sporopollenin: mosses, fern spores and pollen of seed plants have an outer envelope or exine composed of this material, a complex biopolymer resistant to enzymatic degradation and hydrolytic composition.

stalk cell: sterile derivative of the antheridial cell during development of the male gametophyte.

strobilus, strobili (pl.): male or female reproductive structures; composed of a central axis with attached megasporophylls (female) or microsporophylls (male) to which sporangia are attached.

suspensor: a system of cells derived from the upper cells of the proembryo or the S tier of the proembryo.

sulcus: latitudinal opening or aperture on either the distal or the proximal face of the pollen grain.

syngamy: union of two gametes to form a zygote.

tapetum: secretory cells within the microsporangia which form from vegetative (diploid) tissues of the adult sporophyte's microsporophylls; these cells surround and nourish microspore mother cells undergoing meiosis, secrete the exine layer of the pollen wall then degenerate after microspores are released from the tetrad.

tetrad: the product of a single male meiosis is a set of four attached haploid microspores. Each microspore then forms a pollen wall to become a pollen grain.

tetraspory: a rare case where all four meiotic products or megaspores survive then develops into a single haploid female gametophyte.

terminal velocity: an aerodynamic property of pollen which measures the rate at which particles descend in still air owing to gravitational effects.

tube cell: the antheridial initial divides into the antheridial cell and the sterile tube cell during male gametophyte development.

zooidogamy: fertilization by motile male gametes.

Index